T0206032

UNITEXT for Physics

Series editors

Michele Cini, Roma, Italy
Attilio Ferrari, Torino, Italy
Stefano Forte, Milano, Italy
Guido Montagna, Pavia, Italy
Oreste Nicrosini, Pavia, Italy
Luca Peliti, Napoli, Italy
Alberto Rotondi, Pavia, Italy
Paolo Biscari, Milano, Italy
Nicola Manini, Milano, Italy
Morten Hjorth-Jensen, Oslo, Norway

UNITEXT for Physics series, formerly UNITEXT Collana di Fisica e Astronomia, publishes textbooks and monographs in Physics and Astronomy, mainly in English language, characterized of a didactic style and comprehensiveness. The books published in UNITEXT for Physics series are addressed to graduate and advanced graduate students, but also to scientists and researchers as important resources for their education, knowledge and teaching.

More information about this series at http://www.springer.com/series/13351

Luca Salasnich

Quantum Physics of Light and Matter

Photons, Atoms, and Strongly Correlated Systems

Second Edition

 Springer

Luca Salasnich
Fisica e Astronomia "Galileo Galilei"
Università di Padova
Padua
Italy

ISSN 2198-7882   ISSN 2198-7890   (electronic)
UNITEXT for Physics
ISBN 978-3-319-85035-1   ISBN 978-3-319-52998-1   (eBook)
DOI 10.1007/978-3-319-52998-1

© Springer International Publishing AG 2014, 2017
Softcover reprint of the hardcover 2nd edition 2017
This work is subject to copyright. All rights are reserved by the Publisher, whether the whole or part
of the material is concerned, specifically the rights of translation, reprinting, reuse of illustrations,
recitation, broadcasting, reproduction on microfilms or in any other physical way, and transmission
or information storage and retrieval, electronic adaptation, computer software, or by similar or dissimilar
methodology now known or hereafter developed.
The use of general descriptive names, registered names, trademarks, service marks, etc. in this
publication does not imply, even in the absence of a specific statement, that such names are exempt from
the relevant protective laws and regulations and therefore free for general use.
The publisher, the authors and the editors are safe to assume that the advice and information in this
book are believed to be true and accurate at the date of publication. Neither the publisher nor the
authors or the editors give a warranty, express or implied, with respect to the material contained herein or
for any errors or omissions that may have been made. The publisher remains neutral with regard to
jurisdictional claims in published maps and institutional affiliations.

Printed on acid-free paper

This Springer imprint is published by Springer Nature
The registered company is Springer International Publishing AG
The registered company address is: Gewerbestrasse 11, 6330 Cham, Switzerland

Preface to the Second Edition

The second edition of this book has two additional chapters based on lecture notes of the course "Quantum Field Theory in Condensed Matter Physics" given by the author at the Doctoral School in Physics of the University of Padova. These two chapters are an introduction to the functional integration for bosonic and fermionic fields with applications to superfluids, superconductors, and ultracold atomic gases.

Chapter 8 considers the quantum statistical mechanics of bosons within the formalism of functional integration for the bosonic field, introducing the equation of state of weakly interacting bosons and the dimensional regularization of Gaussian fluctuations. The time-dependent Gross-Pitaevskii equation is analyzed: its connection to the equations of superfluid hydrodynamics and its topological and solitonic solutions (quantized vortices, dark solitons, and bright solitons). Chapter 9 introduces the formalism of functional integration for the fermionic field. The equation of state of weakly interacting repulsive fermions and the paring of attractive fermions within the mean-field Bardeen-Cooper-Schrieffer (BCS) theory of low-temperature superconductivity is derived. A section is devoted to the phenomenological Ginzburg-Landau theory of superconductivity which is quite successful near the critical temperature. Finally, the BCS theory is extended to analyze, in the case of ultracold atomic gases, the BCS-BEC crossover from the weak-coupling BCS regime of Cooper pairs to the strong-coupling regime with Bose-Einstein condensation (BEC) of molecules.

The author acknowledges several students who have detected missprints and errors in the first edition. The author thanks Giacomo Bighin, Luca Dell'Anna, Roberto Onofrio, Fabio Sattin, Flavio Toigo, and Giorgio Urso for many suggestions and enlightening discussions about this book.

Padua, Italy
December 2016

Luca Salasnich

Preface to the First Edition

This book contains lecture notes prepared for the one-semester course "Structure of Matter" belonging to the Master of Science in Physics at the University of Padova. The course gives an introduction to the field quantization (second quantization) of light and matter with applications to atomic physics.

The first chapter briefly reviews the origins of special relativity and quantum mechanics and the basic notions of quantum information theory and quantum statistical mechanics. The second chapter is devoted to the second quantization of the electromagnetic field, while the third chapter shows the consequences of the light field quantization in the description of electromagnetic transitions. In the fourth chapter, the spin of the electron, and in particular its derivation from the Dirac equation, is analyzed, while the fifth chapter investigates the effects of external electric and magnetic fields on the atomic spectra (Stark and Zeeman effects). The sixth chapter describes the properties of systems composed by many interacting identical particles. The Fermi degeneracy and the Bose-Einstein condensation, introducing the Hartree-Fock variational method, the density functional theory, and the Born-Oppenheimer approximation, are also discussed. Finally, in the seventh chapter the second quantization of the non-relativistic matter field, i.e., the Schrödinger field, which gives a powerful tool for the investigation of finite-temperature many-body problems and also atomic quantum optics is explained. Moreover, in this last chapter fermionic Fock states and coherent states are presented, and the Hamiltonians of Jaynes-Cummings and Bose-Hubbard are introduced and investigated. Three appendices on the Dirac delta function, the Fourier transform, and the Laplace transform complete the book.

It is important to stress that at the end of each chapter, there are solved problems which help the students to put into practice the things they learned.

Padua, Italy
January 2014

Luca Salasnich

Contents

Chapter 1
The Origins of Modern Physics

In this chapter we review the main results of early modern physics (which we suppose the reader learned in previous introductory courses), namely the special relativity of Einstein and the old quantum mechanics due to Planck, Bohr, Schrödinger, and others. The chapter gives also a brief description of the relevant axioms of quantum mechanics, historically introduced by Dirac and Von Neumann, and elementary notions of quantum information. The chapter ends with elements of quantum statistical mechanics.

1.1 Special Relativity

In 1887 Albert Michelson and Edward Morley made a break-thought experiment of optical interferometry showing that the speed of light in the vacuum is

$$c = 3 \times 10^8 \text{ m/s}, \tag{1.1}$$

independently on the relative motion of the observer (here we have reported an approximated value of c which is correct within three digits). Two years later, Henry Poincaré suggested that the speed of light is the maximum possible value for any kind of velocity. On the basis of previous ideas of George Francis FitzGerald, in 1904 Hendrik Lorentz found that the Maxwell equations of electromagnetism are invariant with respect to this kind of space-time transformations

© Springer International Publishing AG 2017
L. Salasnich, *Quantum Physics of Light and Matter*, UNITEXT for Physics,
DOI 10.1007/978-3-319-52998-1_1

$$x' = \frac{x - vt}{\sqrt{1 - \frac{v^2}{c^2}}} \tag{1.2}$$

$$y' = y \tag{1.3}$$

$$z' = z \tag{1.4}$$

$$t' = \frac{t - vx/c^2}{\sqrt{1 - \frac{v^2}{c^2}}}, \tag{1.5}$$

which are called Lorentz (or Lorentz-FitzGerald) transformations. This research activity on light and invariant transformations was summarized in 1905 by Albert Einstein, who decided to adopt two striking postulates:

 (i) the law of physics are the same for all inertial frames;
(ii) the speed of light in the vacuum is the same in all inertial frames.

From these two postulates Einstein deduced that the laws of physics are invariant with respect to Lorentz transformations but the law of Newtonian mechanics (which are not) must be modified. In this way Einstein developed a new mechanics, the special relativistic mechanics, which reduces to the Newtonian mechanics when the involved velocity v is much smaller than the speed of light c. One of the amazing results of relativistic kinematics is the length contraction: the length L of a rod measured by an observer which moves at velocity v with respect to the rod is given by

$$L = L_0\sqrt{1 - \frac{v^2}{c^2}}, \tag{1.6}$$

where L_0 is the proper length of the rod. Another astonishing result is the time dilatation: the time interval T of a clock measured by an observer which moves at velocity v with respect to the clock is given by

$$T = \frac{T_0}{\sqrt{1 - \frac{v^2}{c^2}}}, \tag{1.7}$$

where T_0 is the proper time interval of the clock.

We conclude this section by observing that, according to the relativistic mechanics of Einstein, the energy E of a particle of rest mass m and linear momentum $p = |\mathbf{p}|$ is given by

$$E = \sqrt{p^2c^2 + m^2c^4}. \tag{1.8}$$

If the particle has zero linear momentum p, i.e. $p = 0$, then

$$E = mc^2, \tag{1.9}$$

which is the rest energy of the particle. Instead, if the linear momentum p is finite but the condition $pc/(mc^2) \ll 1$ holds one can expand the square root finding

$$E = mc^2 + \frac{p^2}{2m} + O(p^4) \,, \tag{1.10}$$

which shows that the energy E is approximated by the sum of two contributions: the rest energy mc^2 and the familiar non-relativistic kinetic energy $p^2/(2m)$. In the case of a particle with zero rest mass, i.e. $m = 0$, the energy is given by

$$E = pc \,, \tag{1.11}$$

which is indeed the energy of light photons. We stress that all the predictions of special relativistic mechanics have been confirmed by experiments. Also many predictions of general relativity, which generalizes the special relativity taking into account the theory of gravitation, have been verified experimentally and also used in applications (for instance, some global positioning system (GPS) devices include general relativity corrections).

1.2 Quantum Mechanics

Historically the beginning of quantum mechanics is set in the year 1900 when Max Planck found that the only way to explain the experimental results of the electromagnetic spectrum emitted by hot solid bodies is to assume that the energy E of the radiation with frequency ν emitted from the walls of the body is quantized according to

$$E = h\nu \, n \,, \tag{1.12}$$

where $n = 0, 1, 2, \ldots$ is an integer quantum number. With the help of statistical mechanics Planck derived the following expression

$$\rho(\nu) = \frac{8\pi^2}{c^3} \nu^2 \frac{h\nu}{e^{\beta h\nu} - 1} \tag{1.13}$$

for the electromagnetic energy density per unit of frequency $\rho(\nu)$ emitted by the hot body at the temperature T, with $k_B = 1.38 \times 10^{-23}$ J/K the Boltzmann constant, $c = 3 \times 10^8$ m/s the speed of light in the vacuum. This formula, known as Planck law of the black-body radiation, is in very good agreement with experimental data. Note that a hot solid body can be indeed approximated by the so-called black body, that is an idealized physical body that absorbs all incident electromagnetic radiation and it is also the best possible emitter of thermal radiation. The constant h derived by Plank from the interpolation of experimental data of radiation reads

$$h = 6.63 \times 10^{-34} \, \text{J s} \,. \tag{1.14}$$

This parameter is called Planck constant. Notice that often one uses instead the reduced Planck constant

$$\hbar = \frac{h}{2\pi} = 1.06 \times 10^{-34} \, \text{J s} \,. \tag{1.15}$$

Few years later the black-body formulation of Planck, in 1905, Albert Einstein suggested that not only the electromagnetic radiation is emitted by hot bodies in a quantized form, as found by Planck, but that indeed electromagnetic radiation is always composed of light quanta, called photons, with discrete energy

$$\epsilon = h\nu \,. \tag{1.16}$$

Einstein used the concept of photon to explain the photoelectric effect and predicted that the kinetic energy of an electron emitted by the surface of a metal after being irradiated is given by

$$\frac{1}{2}mv^2 = h\nu - W \,, \tag{1.17}$$

where W is work function of the metal (i.e. the minimum energy to extract the electron from the surface of the metal), m is the mass of the electron, and v is the emission velocity of the electron. This prediction clearly implies that the minimal radiation frequency to extract electrons from a metal is $\nu_{min} = W/h$. Subsequent experiments confirmed the Einstein's formula and gave a complementary measure of the Planck constant h.

In 1913 Niels Bohr was able to explain the discrete frequencies of electromagnetic emission of hydrogen atom under the hypothesis that the energy of the electron orbiting around the nucleus is quantized according to the formula

$$E_n = -\frac{me^4}{2\varepsilon_0^2 h^2} \frac{1}{n^2} = -13.6 \, \text{eV} \, \frac{1}{n^2} \,, \tag{1.18}$$

where $n = 1, 2, 3, \ldots$ is the principal integer quantum number, e is the electric charge of the electron and ε_0 the dielectric constant of the vacuum. This expression shows that the quantum states of the system are characterized by the quantum number n and the ground-state ($n = 1$) has the energy -13.6 eV, which is the ionization energy of the hydrogen. According to the theory of Bohr, the electromagnetic radiation is emitted or absorbed when one electron has a transition from one energy level E_n to another E_m. In addition, the frequency ν of the radiation is related to the energies of the two states involved in the transition by

$$h\nu = E_n - E_m \,. \tag{1.19}$$

Thus any electromagnetic transition between two quantum states implies the emission or the absorption of one photon with an energy $h\nu$ equal to the energy difference of the involved states.

In 1922 Arthur Compton noted that the diffusion of X rays with electrons (Compton effect) is a process of scattering between photons of X rays and electrons. The photon of X rays has the familiar energy

$$\epsilon = h\nu = h\frac{c}{\lambda} , \tag{1.20}$$

with λ wavelength of the photon, because the speed of light is given by $c = \lambda\nu$. By using the connection relativistic between energy ϵ and linear momentum $p = |\mathbf{p}|$ for a particle with zero rest mass, Compton found

$$\epsilon = pc , \tag{1.21}$$

where the linear momentum of the photon can be then written as

$$p = \frac{h}{\lambda} . \tag{1.22}$$

By applying the conservation of energy and linear momentum to the scattering process Compton got the following expression

$$\lambda' - \lambda = \frac{h}{mc} (1 - \cos(\theta)) \tag{1.23}$$

for the wavelength λ' of the diffused photon, with θ the angle between the incoming photon and the outcoming one. This formula is in full agreement with experimental results.

Inspired by the wave-particle behavior exhibited by the light, in 1924 Louis de Broglie suggested that also the matter, and in particular the electron, has wave-like properties. He postulated that the relationship

$$\lambda = \frac{h}{p} \tag{1.24}$$

applies not only to photons but also to material particles. In general, p is the linear momentum (particle property) and λ the wavelength (wave property) of the quantum entity, that is usually called quantum particle.

Two years later the proposal of de Broglie, in 1926, Erwin Schrödinger went to the extremes of the wave-particle duality and introduced the following wave equation for one electron under the action of an external potential $U(\mathbf{r})$

$$i\hbar\frac{\partial}{\partial t}\psi(\mathbf{r}, t) = \left[-\frac{\hbar^2}{2m}\nabla^2 + U(\mathbf{r})\right]\psi(\mathbf{r}, t) , \tag{1.25}$$

where $\nabla^2 = \frac{\partial^2}{\partial x^2} + \frac{\partial^2}{\partial y^2} + \frac{\partial^2}{\partial z^2}$ is the Laplacian operator. This equation is now known as the Schrödinger equation. In the case of the hydrogen atom, where

$$U(\mathbf{r}) = -\frac{e^2}{4\pi\varepsilon_0 \, r^2} \, , \qquad (1.26)$$

Schrödinger showed that setting

$$\psi(\mathbf{r}, t) = R_n(r) \, e^{-iE_n t/\hbar} \qquad (1.27)$$

one finds

$$\left[-\frac{\hbar^2}{2m}\nabla^2 - \frac{e^2}{4\pi\varepsilon_0 \, r^2} \right] R_n(r) = E_n R_n(r) \qquad (1.28)$$

the stationary Schrödinger equation for the radial eigenfunction $R_n(r)$ with eigenvalue E_n given exactly by the Bohr formula (1.18). Soon after, it was understood that the stationary Schrödinger equation of the hydrogen atom satisfies a more general eigenvalue problem, namely

$$\left[-\frac{\hbar^2}{2m}\nabla^2 - \frac{e^2}{4\pi\varepsilon_0 \, r^2} \right] \psi_{nlm_l}(\mathbf{r}) = E_n \psi_{nlm_l}(\mathbf{r}) \, , \qquad (1.29)$$

where $\psi_{nlm_l}(\mathbf{r})$ is a generic eigenfunction of the problem, which depends on three quantum numbers: the principal quantum number $n = 1, 2, 3, \ldots$, the angular quantum number $l = 0, 1, 2, \ldots, n - 1$, and third-component angular quantum number $m_l = -l, -l + 1, \ldots, l - 1, l$. The generic eigenfunction of the electron in the hydrogen atom in spherical coordinates is given by

$$\psi_{nlm_l}(r, \theta, \phi) = R_{nl}(r) \, Y_{lm_l}(\theta, \phi) \, ,$$

where $R_{nl}(r)$ is the radial wavefunction while $Y_{lm_l}(\theta, \phi)$ is the angular wavefunction. Notice that in Eq. (1.28) we have $R_n(r) = R_{n0}(r)$. It is important to stress that initially Schrödinger thought that $\psi(\mathbf{r})$ was a matter wave, such that $|\psi(\mathbf{r}, t)|^2$ gives the local density of electrons in the position $\mathbf{r}$ at time t. It was Max Born that correctly interpreted $\psi(\mathbf{r}, t)$ as a probability field, where $|\psi(\mathbf{r}, t)|^2$ is the local probability density of finding one electron in the position $\mathbf{r}$ at time t, with the normalization condition

$$\int d^3\mathbf{r} \, |\psi(\mathbf{r}, t)|^2 = 1 \, . \qquad (1.30)$$

In the case of N particles the probabilistic interpretation of Born becomes crucial, with $\Psi(\mathbf{r}_1, \mathbf{r}_2, \ldots, \mathbf{r}_N, t)$ the many-body wavefunction of the system, such that $|\Psi(\mathbf{r}_1, \mathbf{r}_2, \ldots, \mathbf{r}_N, t)|^2$ is the probability of finding at time t one particle in the position $\mathbf{r}_1$, another particle in the position $\mathbf{r}_2$, and so on.

In the same year of the discovery of the Schrödinger equation Max Born, Pasqual Jordan and Werner Heisenberg introduced the matrix mechanics. According to this theory the position $\mathbf{r}$ and the linear momentum $\mathbf{p}$ of an elementary particle are not vectors composed of numbers but instead vector composed of matrices (operators) which satisfies strange commutation rules, i.e.

$$\hat{\mathbf{r}} = (\hat{x}, \hat{y}, \hat{z}) , \quad \hat{\mathbf{p}} = (\hat{p}_x, \hat{p}_y, \hat{p}_z) , \tag{1.31}$$

such that

$$[\hat{\mathbf{r}}, \hat{\mathbf{p}}] = i\hbar , \tag{1.32}$$

where the hat symbol is introduced to denote operators and $[\hat{A}, \hat{B}] = \hat{A}\hat{B} - \hat{B}\hat{A}$ is the commutator of generic operators $\hat{A}$ and $\hat{B}$. By using their theory Born, Jordan and Heisenberg were able to obtain the energy spectrum of the hydrogen atom and also to calculate the transition probabilities between two energy levels. Soon after, Schrödinger realized that the matrix mechanics is equivalent to his wave-like formulation: introducing the quantization rules

$$\hat{\mathbf{r}} = \mathbf{r} , \quad \hat{\mathbf{p}} = -i\hbar\nabla , \tag{1.33}$$

for any function $f(\mathbf{r})$ one finds immediately the commutation rule

$$\left(\hat{\mathbf{r}} \cdot \hat{\mathbf{p}} - \hat{\mathbf{p}} \cdot \hat{\mathbf{r}}\right) f(\mathbf{r}) = (-i\hbar\mathbf{r} \cdot \nabla + i\hbar\nabla \cdot \mathbf{r}) f(\mathbf{r}) = i\hbar f(\mathbf{r}) . \tag{1.34}$$

Moreover, starting from classical Hamiltonian

$$H = \frac{p^2}{2m} + U(\mathbf{r}) , \tag{1.35}$$

the quantization rules give immediately the quantum Hamiltonian operator

$$H = -\frac{\hbar^2}{2m}\nabla^2 + U(\mathbf{r}) \tag{1.36}$$

from which one can write the time-dependent Schrödinger equation as

$$i\hbar\frac{\partial}{\partial t}\psi(\mathbf{r}, t) = \hat{H}\psi(\mathbf{r}, t) . \tag{1.37}$$

In 1927 the wave-like behavior of electrons was eventually demonstrated by Clinton Davisson and Lester Germer, who observed the diffraction of a beam of electrons across a solid crystal. The diffracted beam shows intensity maxima when the following relationship is satisfied

$$2\,d\,\sin(\phi) = n\,\lambda , \tag{1.38}$$

where n is an integer number, λ is the de Broglie wavelength of electrons, d is the separation distance of crystal planes and ϕ is the angle between the incident beam of electrons and the surface of the solid crystal. The condition for a maximum diffracted beam observed in this experiment corresponds indeed to the constructive interference of waves, which is well know in optics (Bragg condition).

1.2.1   Axioms of Quantum Mechanics

The axiomatic formulation of quantum mechanics was set up by Paul Maurice Dirac in 1930 and John von Neumann in 1932. The basic axioms are the following:

Axiom 1. The state of a quantum system is described by a unitary vector $|\psi\rangle$ belonging to a separable complex Hilbert space.

Axiom 2. Any observable of a quantum system is described by a self-adjoint linear operator $\hat{F}$ acting on the Hilbert space of state vectors.

Axiom 3. The possible measurable values of an observable $\hat{F}$ are its eigenvalues f, such that

$$\hat{F}|f\rangle = f|f\rangle$$

with $|f\rangle$ the corresponding eigenstate. Note that the observable $\hat{F}$ admits the spectral resolution

$$\hat{F} = \sum_f |f\rangle f \langle f| \, ,$$

which is quite useful in applications, as well as the spectral resolution of the identity

$$\hat{I} = \sum_f |f\rangle \langle f| \, .$$

Axiom 4. The probability p of finding the state $|\psi\rangle$ in the state $|f\rangle$ is given by

$$p = |\langle f|\psi\rangle|^2 \, ,$$

where the complex probability amplitude $\langle f|\psi\rangle$ denotes the scalar product of the two vectors. This probability p is also the probability of measuring the value f of the observable $\hat{F}$ when the system in the quantum state $|\psi\rangle$. Notice that both $|\psi\rangle$ and $|f\rangle$ must be normalized to one. Often it is useful to introduce the expectation value (mean value or average value) of an observable $\hat{F}$ with respect to a state $|\psi\rangle$ as $\langle\psi|\hat{F}|\psi\rangle$. Moreover, from this Axiom 4 it follows that the wavefunction $\psi(\mathbf{r})$ can be interpreted as

$$\psi(\mathbf{r}) = \langle \mathbf{r}|\psi\rangle \, ,$$

that is the probability amplitude of finding the state $|\psi\rangle$ in the position state $|\mathbf{r}\rangle$.

Axiom 5. The time evolution of states and observables of a quantum system with Hamiltonian $\hat{H}$ is determined by the unitary operator

$$\hat{U}(t) = \exp\left(-i\hat{H}t/\hbar\right),$$

such that $|\psi(t)\rangle = \hat{U}(t)|\psi\rangle$ is the time-evolved state $|\psi\rangle$ at time t and $\hat{F}(t) = \hat{U}^{-1}(t)\hat{F}\hat{U}(t)$ is the time-evolved observable $\hat{F}$ at time t. From this axiom one finds immediately the Schrödinger equation

$$i\hbar\frac{\partial}{\partial t}|\psi(t)\rangle = \hat{H}|\psi(t)\rangle$$

for the state $|\psi(t)\rangle$, and the Heisenberg equation

$$i\hbar\frac{\partial}{\partial t}\hat{F}(t) = [\hat{F}(t), \hat{H}]$$

for the observable $\hat{F}(t)$.

1.2.2 Quantum Information

The qubit, or quantum bit, is the quantum analogue of a classical bit. It is the unit of quantrum information, namely a tw-level quantum system. There are many two-level system which can be used as a physical realization of the qubit. For instance: horizontal and vertical polarizations of light, ground and excited states of atoms or molecules or nuclei, left and right wells of a double-well potential, up and down spins of a particle.

The two basis states of the qubit are usually denoted as $|0\rangle$ and $|1\rangle$. They can be written as

$$|0\rangle = \begin{pmatrix} 1 \\ 0 \end{pmatrix}, \quad |1\rangle = \begin{pmatrix} 0 \\ 1 \end{pmatrix}, \tag{1.39}$$

by using a vector representation, where clearly

$$\langle 0| = (1, 0), \quad \langle 1| = (0, 1), \tag{1.40}$$

and also

$$\langle 0|0\rangle = (1, 0)\begin{pmatrix} 1 \\ 0 \end{pmatrix} = 1, \tag{1.41}$$

$$\langle 0|1\rangle = (1, 0)\begin{pmatrix} 0 \\ 1 \end{pmatrix} = 0, \tag{1.42}$$

$$\langle 1|0 \rangle = (0, 1) \begin{pmatrix} 1 \\ 0 \end{pmatrix} = 0 , \tag{1.43}$$

$$\langle 1|1 \rangle = (0, 1) \begin{pmatrix} 0 \\ 1 \end{pmatrix} = 1 , \tag{1.44}$$

while

$$|0\rangle\langle 0| = \begin{pmatrix} 1 \\ 0 \end{pmatrix} (1, 0) = \begin{pmatrix} 1 & 0 \\ 0 & 0 \end{pmatrix} , \tag{1.45}$$

$$|0\rangle\langle 1| = \begin{pmatrix} 1 \\ 0 \end{pmatrix} (0, 1) = \begin{pmatrix} 0 & 1 \\ 0 & 0 \end{pmatrix} , \tag{1.46}$$

$$|1\rangle\langle 0| = \begin{pmatrix} 0 \\ 1 \end{pmatrix} (1, 0) = \begin{pmatrix} 0 & 0 \\ 1 & 0 \end{pmatrix} , \tag{1.47}$$

$$|1\rangle\langle 1| = \begin{pmatrix} 0 \\ 1 \end{pmatrix} (0, 1) = \begin{pmatrix} 0 & 0 \\ 0 & 1 \end{pmatrix} . \tag{1.48}$$

Of course, instead of $|0\rangle$ and $|1\rangle$ one can choose other basis states. For instance:

$$|+\rangle = \frac{1}{\sqrt{2}} (|0\rangle + |1\rangle) = \frac{1}{\sqrt{2}} \begin{pmatrix} 1 \\ 1 \end{pmatrix} , \qquad |-\rangle = \frac{1}{\sqrt{2}} (|0\rangle - |1\rangle) = \frac{1}{\sqrt{2}} \begin{pmatrix} 1 \\ -1 \end{pmatrix} , \tag{1.49}$$

but also

$$|i\rangle = \frac{1}{\sqrt{2}} (|0\rangle + i|1\rangle) = \frac{1}{\sqrt{2}} \begin{pmatrix} 1 \\ i \end{pmatrix} , \qquad |-i\rangle = \frac{1}{\sqrt{2}} (|0\rangle - i|1\rangle) = \frac{1}{\sqrt{2}} \begin{pmatrix} 1 \\ -i \end{pmatrix} . \tag{1.50}$$

A pure qubit $|\psi\rangle$ is a linear superposition (superposition state) of the basis states $|0\rangle$ and $|1\rangle$, i.e.

$$|\psi\rangle = \alpha\, |0\rangle + \beta\, |1\rangle , \tag{1.51}$$

where α and β are the probability amplitudes, usually complex numbers, such that

$$|\alpha|^2 + |\beta|^2 = 1 . \tag{1.52}$$

A quantum gate (or quantum logic gate) is a unitary operator $\hat{U}$ acting on qubits. Among the quantum gates acting on a single qubit there are: the identity gate

$$\hat{I} = \begin{pmatrix} 1 & 0 \\ 0 & 1 \end{pmatrix} , \tag{1.53}$$

the Hadamard gate

$$\hat{H} = \frac{1}{\sqrt{2}} \begin{pmatrix} 1 & 1 \\ 1 & -1 \end{pmatrix} , \tag{1.54}$$

the not gate (also called Pauli X gate)

$$\hat{X} = \begin{pmatrix} 0 & 1 \\ 1 & 0 \end{pmatrix} = \hat{\sigma}_1 , \tag{1.55}$$

the Pauli Y gate

$$\hat{Y} = \begin{pmatrix} 0 & -i \\ i & 0 \end{pmatrix} = \hat{\sigma}_2 , \tag{1.56}$$

and the Pauli Z gate

$$\hat{Z} = \begin{pmatrix} 1 & 0 \\ 0 & -1 \end{pmatrix} = \hat{\sigma}_3 . \tag{1.57}$$

It is strightforward to deduce the properties of the quantum gates. For example, one easily finds

$$\hat{H}|0\rangle = |+\rangle , \qquad \hat{H}|1\rangle = |-\rangle , \tag{1.58}$$

but also

$$\hat{H}|+\rangle = |0\rangle , \qquad \hat{H}|-\rangle = |1\rangle . \tag{1.59}$$

A N-qubit $|\Phi_N\rangle$, also called a quantum register, is a quantum state characterized by

$$|\Phi_N\rangle = \sum_{\alpha_1...\alpha_N} c_{\alpha_1...\alpha_N} |\alpha_1\rangle \otimes ... \otimes |\alpha_N\rangle , \tag{1.60}$$

where $|\alpha_i\rangle$ is a single qubit (with $\alpha_i = 0, 1$), and

$$\sum_{\alpha_1...\alpha_N} |c_{\alpha_1...\alpha_N}|^2 = 1 . \tag{1.61}$$

In practice, the N-qubit describes N two-state configurations. The more general 2-qubit $|\Phi_2\rangle$ is consequently given by

$$|\Phi_2\rangle = c_{00}|00\rangle + c_{01}|01\rangle + c_{10}|10\rangle + c_{11}|11\rangle , \tag{1.62}$$

where we use $|00\rangle = |0\rangle \otimes |0\rangle, |01\rangle = |0\rangle \otimes |1\rangle, |10\rangle = |1\rangle \otimes |0\rangle$, and $|11\rangle = |1\rangle \otimes |1\rangle$ to simplify the notation.

For the basis states of 2-qubits we one introduce the following vector representation

$$|00\rangle = \begin{pmatrix} 1 \\ 0 \\ 0 \\ 0 \end{pmatrix} , \quad |01\rangle = \begin{pmatrix} 0 \\ 1 \\ 0 \\ 0 \end{pmatrix} , \quad |10\rangle = \begin{pmatrix} 0 \\ 0 \\ 1 \\ 0 \end{pmatrix} , \quad |11\rangle = \begin{pmatrix} 0 \\ 0 \\ 0 \\ 1 \end{pmatrix} . \tag{1.63}$$

Moreover, quantum gates can be introduced also for 2-qubits. Among the quantum gates acting on a 2-qubit there are: the identity gate

$$\hat{I} = \begin{pmatrix} 1 & 0 & 0 & 0 \\ 0 & 1 & 0 & 0 \\ 0 & 0 & 1 & 0 \\ 0 & 0 & 0 & 1 \end{pmatrix} \tag{1.64}$$

and the CNOT (also called controlled-NOT) gate

$$C\hat{N}OT = \begin{pmatrix} 1 & 0 & 0 & 0 \\ 0 & 1 & 0 & 0 \\ 0 & 0 & 0 & 1 \\ 0 & 0 & 1 & 0 \end{pmatrix} . \tag{1.65}$$

Notice that the action of the CNOT gate on a state $|\alpha\beta\rangle = |\alpha\rangle|\beta\rangle$ is such that the "target" qubit $|\beta\rangle$ changes only if the "control" qubit $|\alpha\rangle$ is $|1\rangle$.

The 2-qubit $|\Phi_2\rangle$ is called separable if it can be written as the tensor product of two generic single qubits $|\psi_A\rangle$ and $|\psi_B\rangle$, i.e.

$$|\Phi_2\rangle = |\psi_A\rangle \otimes |\psi_B\rangle . \tag{1.66}$$

If this is not possible, the state $|\Phi_2\rangle$ is called entangled. Examples of separable states are

$$|00\rangle = |0\rangle \otimes |0\rangle , \tag{1.67}$$

or

$$|01\rangle = |0\rangle \otimes |1\rangle , \tag{1.68}$$

but also

$$\frac{1}{\sqrt{2}} \left(|01\rangle + |11\rangle \right) = \frac{1}{\sqrt{2}} \left(|0\rangle + |1\rangle \right) \otimes |1\rangle . \tag{1.69}$$

Examples of entangled states are instead

$$\frac{1}{\sqrt{2}} \left(|01\rangle \pm |10\rangle \right) . \tag{1.70}$$

but also

$$\frac{1}{\sqrt{2}} \left(|00\rangle \pm |11\rangle \right) , \tag{1.71}$$

which are called Bell states.

Remarkably, a CNOT gate acting on a separable state can produce an entagled state, and viceversa. For instance:

$$C\hat{N}OT\frac{1}{\sqrt{2}}(|00\rangle + |10\rangle) = \frac{1}{\sqrt{2}}(|00\rangle + |11\rangle) , \qquad (1.72)$$

while

$$C\hat{N}OT\frac{1}{\sqrt{2}}(|00\rangle + |11\rangle) = \frac{1}{\sqrt{2}}(|00\rangle + |10\rangle) . \qquad (1.73)$$

1.3 Quantum Statistical Mechanics

Statistical mechanics aims to describe macroscopic properties of complex systems starting from their microscopic components by using statistical averages. Statistical mechanics must reproduce the general results of Thermodynamics both at equilibrium and out of equilibrium. Clearly, the problem is strongly simplified if the system is at thermal equilibrium. Here we discuss only quantum systems at thermal equilibrium and consider a many-body quantum system of identical particles characterized by the Hamiltonian $\hat{H}$ such that

$$\hat{H}|E_i^{(N)}\rangle = E_i^{(N)}|E_i^{(N)}\rangle , \qquad (1.74)$$

where $|E_i^{(N)}\rangle$ are the eigenstates of $\hat{H}$ for a fixed number N of identical particles and E_i^N are the corresponding eigenenergies.

1.3.1 Microcanonical Ensemble

In the microcanonical ensemble the quantum many-body system in a volume V has a fixed number N of particles and also a fixed energy E. In this case the Hamiltonian $\hat{H}$ admits the spectral decomposition

$$\hat{H} = \sum_i E_i^{(N)}|E_i^{(N)}\rangle\langle E_i^{(N)}| , \qquad (1.75)$$

and one defines the microcanonical density operator as

$$\hat{\rho} = \delta(E - \hat{H}) , \qquad (1.76)$$

where $\delta(x)$ is the Dirac delta function. This microcanonical density operator $\hat{\rho}$ has the spectral decomposition

$$\hat{\rho} = \sum_i \delta(E - E_i^{(N)})|E_i^{(N)}\rangle\langle E_i^{(N)}| . \qquad (1.77)$$

The key quantity in the microcanonical ensemble is the density of states (or micro-canonical volume) W given by

$$W = Tr[\hat{\rho}] = Tr[\delta(E - \hat{H})] , \qquad (1.78)$$

namely

$$W = \sum_i \delta(E - E_i^{(N)}) . \qquad (1.79)$$

The ensemble average of an observable described by the self-adjunct operator $\hat{A}$ is defined as

$$\langle \hat{A} \rangle = \frac{Tr[\hat{A}\,\hat{\rho}]}{Tr[\hat{\rho}]} = \frac{1}{W} \sum_i A_{ii}^{(N)}\, \delta(E - E_i^{(N)}) , \qquad (1.80)$$

where $A_{ii}^{(N)} = \langle E_i^{(N)}|\hat{A}|E_i^{(N)}\rangle$. The connection with Equilibrium Thermodynamics is given by the formula

$$S = k_B \ln(W) , \qquad (1.81)$$

which introduces the entropy S as a function of energy E, volume V and number N of particles. Note that Eq. (1.81) was discovered by Ludwig Boltzmann in 1872 and $k_B = 1.38 \times 10^{-23}$ J/K is the Boltzmann constant. From the entropy $S(E, V, N)$ the absolute temperature T, the pressure P and the chemical potential μ are obtained as

$$\frac{1}{T} = \left(\frac{\partial S}{\partial E}\right)_{V,N} , \qquad P = T\left(\frac{\partial S}{\partial V}\right)_{E,N} , \qquad \mu = -T\left(\frac{\partial S}{\partial N}\right)_{E,V} , \qquad (1.82)$$

which are familiar relationships of Equilibrium Thermodynamics such that

$$dS = \frac{1}{T}dE + \frac{P}{T}dV - \frac{\mu}{T}dN . \qquad (1.83)$$

In fact, in the microcanonical ensemble the independent thermodynamic variables are E, N and V, while T, P and μ are dependent thermodynamic variables.

1.3.2 Canonical Ensemble

In the canonical ensemble the quantum system in a volume V has a fixed number N of particles and a fixed temperature T. In this case one defines the canonical density operator as

$$\hat{\rho} = e^{-\beta\hat{H}} , \qquad (1.84)$$

with $\beta = 1/(k_B T)$. Here $\hat{\rho}$ has the spectral decomposition

$$\hat{\rho} = \sum_i e^{-\beta E_i^{(N)}} |E_i^{(N)}\rangle \langle E_i^{(N)}| \ . \tag{1.85}$$

The key quantity in the canonical ensemble is the canonical partition function (or canonical volume) $\mathcal{Z}_N$ given by

$$\mathcal{Z}_N = Tr[\hat{\rho}] = Tr[e^{-\beta \hat{H}}] \ , \tag{1.86}$$

namely

$$\mathcal{Z}_N = \sum_i e^{-\beta E_i^{(N)}} \ . \tag{1.87}$$

The ensemble average of an observable $\hat{A}$ is defined as

$$\langle \hat{A} \rangle = \frac{Tr[\hat{A}\,\hat{\rho}]}{Tr[\hat{\rho}]} = \frac{1}{\mathcal{Z}_N} \sum_i A_{ii}^{(N)} \, e^{-\beta E_i^{(N)}} \ , \tag{1.88}$$

where $A_{ii}^{(N)} = \langle E_i^{(N)}|\hat{A}|E_i^{(N)}\rangle$. Notice that the definition of canonical-ensemble average is the same of the microcanonical-ensemble average but the density of state $\hat{\rho}$ is different in the two ensembles. The connection with Equilibrium Thermodynamics is given by the formula

$$\mathcal{Z}_N = e^{-\beta F} \ , \tag{1.89}$$

which introduces the Helmholtz free energy F as a function of temperature T, volume V and number N of particles. From the Helholtz free energy $F(T, V, N)$ the entropy S, the pressure P and the chemical potential μ are obtained as

$$S = -\left(\frac{\partial F}{\partial T}\right)_{V,N} \ , \qquad P = -\left(\frac{\partial F}{\partial V}\right)_{T,N} \ , \qquad \mu = \left(\frac{\partial F}{\partial N}\right)_{T,V} \ , \tag{1.90}$$

which are familiar relationships of Equilibrium Thermodynamics such that

$$dF = -SdT - PdV + \mu dN \ . \tag{1.91}$$

In fact, in the canonical ensemble the independent thermodynamic variables are T, N and V, while S, P and μ are dependent thermodynamic variables.

1.3.3 Grand Canonical Ensemble

In the grand canonical ensemble the quantum system in a volume V has a fixed temperature T and a fixed chemical potential μ. In this case the Hamiltonian $\hat{H}$ has the spectral decomposition

$$\hat{H} = \sum_{N=0}^{\infty} \sum_{i} E_i^{(N)} |E_i^{(N)}\rangle \langle E_i^{(N)}| , \tag{1.92}$$

which is a generalization of Eq. (1.75), and one introduces the total number operator $\hat{N}$ such that

$$\hat{N} |E_i^{(N)}\rangle = N |E_i^{(N)}\rangle , \tag{1.93}$$

and consequently $\hat{N}$ has the spectral decomposition

$$\hat{N} = \sum_{N=0}^{\infty} \sum_{i} N |E_i^{(N)}\rangle \langle E_i^{(N)}| . \tag{1.94}$$

For the grand canonical ensemble one defines the grand canonical density operator as

$$\hat{\rho} = e^{-\beta(\hat{H} - \mu\hat{N})} , \tag{1.95}$$

with $\beta = 1/(k_B T)$ and μ the chemical potential. Here $\hat{\rho}$ has the spectral decomposition

$$\hat{\rho} = \sum_{N=0}^{\infty} \sum_{i} e^{-\beta(E_i^{(N)} - \mu N)} |E_i^{(N)}\rangle \langle E_i^{(N)}| = \sum_{N=0}^{\infty} z^N \sum_{i} e^{-\beta E_i^{(N)}} |E_i^{(N)}\rangle \langle E_i^{(N)}| , \tag{1.96}$$

where $z = e^{\beta\mu}$ is the fugacity. The key quantity in the grand canonical ensemble is the grand canonical partition function (or grand canonical volume) $\mathcal{Z}$ given by

$$\mathcal{Z} = Tr[\hat{\rho}] = Tr[e^{-\beta(\hat{H} - \mu\hat{N})}] , \tag{1.97}$$

namely

$$\mathcal{Z} = \sum_{N=0}^{\infty} \sum_{i} e^{-\beta(E_i^{(N)} - \mu N)} = \sum_{N=0}^{\infty} z^N \mathcal{Z}_N . \tag{1.98}$$

The ensemble average of an observable $\hat{A}$ is defined as

$$\langle \hat{A} \rangle = \frac{Tr[\hat{A} \hat{\rho}]}{Tr[\hat{\rho}]} = \frac{1}{\mathcal{Z}} \sum_{N=0}^{\infty} \sum_{i} A_{ii}^{(N)} e^{-(\beta E_i^{(N)} - \mu N)} , \tag{1.99}$$

where $A_{ii}^{(N)} = \langle E_i^{(N)} | \hat{A} | E_i^{(N)} \rangle$. Notice that the definition of grand canonical-ensemble average is the same of both microcanonical-ensemble average and canonical-ensemble average but the density of state $\hat{\rho}$ is different in the three ensembles. The connection with Equilibrium Thermodynamics is given by the formula

$$\mathcal{Z} = e^{-\beta\Omega} , \tag{1.100}$$

which introduces the grand potential Ω as a function of temperature T, volume V and chemical potential μ. From the grand potential $\Omega(T, V, \mu)$ the entropy S, the pressure P and the average number $\bar{N}$ of particles are obtained as

$$S = -\left(\frac{\partial \Omega}{\partial T}\right)_{V,\mu}, \quad P = -\left(\frac{\partial \Omega}{\partial V}\right)_{T,\mu}, \quad \bar{N} = -\left(\frac{\partial \Omega}{\partial \mu}\right)_{V,T}, \quad (1.101)$$

which are familiar relationships of Equilibrium Thermodynamics such that

$$d\Omega = -S\,dT - P\,dV - \bar{N}\,d\mu. \quad (1.102)$$

In fact, in the grand canonical ensemble the independent thermodynamic variables are T, μ and V, while S, P and $\bar{N}$ are dependent thermodynamic variables.

To conclude this section we observe that in the grand canonical ensemble, instead of working with eigenstates $|E_i^{(N)}\rangle$ of $\hat{H}$ at fixed number N of particles, one can work with multi-mode Fock states

$$|n_0 \, n_1 \, n_2 \, \ldots \, n_\infty\rangle = |n_0\rangle \otimes |n_1\rangle \otimes |n_2\rangle \otimes \ldots \otimes |n_\infty\rangle, \quad (1.103)$$

where $|n_\alpha\rangle$ is the single-mode Fock state which describes n_α particles in the single-mode state $|\alpha\rangle$ with $\alpha = 0, 1, 2, \ldots$. The trace Tr which appears in Eq. (1.97) is indeed independent on the basis representation. We analyze this issue in Chaps. 2 and 7.

1.4 Solved Problems

Problem 1.1
The electromagnetic radiation of the black body has the following energy density per unit of frequency

$$\rho(\nu) = \frac{8\pi^2}{c^3}\nu^2\frac{h\nu}{e^{\beta h\nu} - 1},$$

where c is the speed of light in vacuum, h is the Planck constant, and $\beta = 1/(k_B T)$ with T absolute temperature and k_B Boltzmann constant. Determine the corresponding energy density per unit of wavelength.

Solution
The energy density $\rho(\nu)$ is such that

$$\mathcal{E} = \int_0^\infty \rho(\nu)\,d\nu$$

represents the energy of the radiation per unit of volume, i.e. the energy density. The linear frequency ν is related to the wavelength λ by the expression

$$\lambda \nu = c .$$

In practice, $\nu = c\lambda^{-1}$, from which

$$d\nu = -c\lambda^{-2}d\lambda .$$

By changing variable the energy density becomes

$$\mathcal{E} = \int_0^\infty \frac{8\pi^2}{\lambda^5} \frac{hc}{e^{\beta hc/\lambda} - 1} d\lambda ,$$

and consequently the energy density per unit of wavelength reads

$$\rho(\lambda) = \frac{8\pi^2}{\lambda^5} \frac{hc}{e^{\beta hc/\lambda} - 1} ,$$

such that

$$\mathcal{E} = \int_0^\infty \rho(\lambda) \, d\lambda .$$

Problem 1.2
Calculate the number of photon emitted in 4 s by a lamp of 10 W which radiates 1% of its energy as monochromatic light with wavelength 6000×10^{-10} m (orange light).

Solution
The energy of one photon of wavelength λ and linear frequency ν is given by

$$\epsilon = h\nu = h\frac{c}{\lambda} ,$$

where

$$h = 6.63 \times 10^{-34} \text{ J s}$$

is the Planck constant. In our problem one gets

$$\epsilon = h\frac{c}{\lambda} = 6.62 \times 10^{-34} \text{ J s} \frac{3 \times 10^8 \text{ m/s}}{6 \times 10^3 \times 10^{-10} \text{ m}} = 3.3 \times 10^{-19} \text{ J} .$$

During the period $\Delta t = 4$ s the energy of the lamp with power $P = 10$ W is

$$E = P \, \Delta t = 10 \text{ J/s} \times 4 \text{ s} = 40 \text{ J} .$$

The radiation energy is instead

$$E_{rad} = E \times 1\% = E \times \frac{1}{100} = \frac{40}{100} \, \text{J} = 0.4 \, \text{J} \,.$$

The number of emitted photons is then

$$N = \frac{E_{rad}}{\epsilon} = \frac{0.4 \, \text{J}}{3.3 \times 10^{-19} \, \text{J}} = 1.2 \times 10^{18} \,.$$

Problem 1.3
On a photoelectric cell it arrives a beam of light with wavelength 6500×10^{-10} m and energy 10^6 erg per second [1 erg = 10^{-7} J]. This energy is entirely used to produce photoelectrons. Calculate the intensity of the electric current which flows in the electric circuit connected to the photoelectric cell.

Solution
The wavelength can be written as

$$\lambda = 6.5 \times 10^3 \times 10^{-10} \, \text{m} = 6.5 \times 10^{-7} \, \text{m} \,.$$

The energy of the beam of light can be written as

$$E = 10^6 \, erg = 10^6 \times 10^{-7} \, \text{J} = 10^{-1} \text{J} \,.$$

The time interval is
$$\Delta t = 1 \, \text{s} \,.$$

The energy of a single photon reads

$$\epsilon = h\frac{c}{\lambda} = 6.6 \times 10^{-34} \frac{3 \times 10^8}{6.5 \times 10^{-7}} \, \text{J} = 3 \times 10^{-19} \, \text{J} \,,$$

where h is the Planck constant and c the speed of light in the vacuum.
The number of photons is thus given by

$$N = \frac{E}{\epsilon} = \frac{10^{-1} \, \text{J}}{3 \times 10^{-19} \, \text{J}} = 3.3 \times 10^{17} \,.$$

If one photon produces one electron with electric charge $e = -1.6 \times 10^{-19}$ C, the intensity of electric current is easily obtained:

$$I = \frac{|e|N}{\Delta t} = \frac{1.6 \times 10^{-19} \, \text{C} \times 3.3 \times 10^{17}}{1 \, \text{s}} = 5.5. \times 10^{-2} \, \text{A} \,.$$

To conclude we observe that with wavelength 6500×10^{-10} m the photoelectric effect is possible only if the work function of the sample is reduced, for instance by using an external electric field.

Problem 1.4

Prove that the simple 2-qubit state

$$|\Phi_2\rangle = c_{00}|01\rangle + c_{10}|10\rangle$$

is separable if and only if $c_{01} = 0$ or $c_{10} = 0$.

Solution

Let us consider two generic 1-qubits, given by

$$|\psi_A\rangle = \alpha_A|0\rangle + \beta_A|1\rangle , \qquad |\psi_B\rangle = \alpha_B|0\rangle + \beta_B|1\rangle .$$

Then

$$|\psi_A\rangle|\psi_B\rangle = \alpha_A\alpha_B|00\rangle + \alpha_A\beta_B|01\rangle + \beta_A\alpha_B|10\rangle + \beta_A\beta_B|11\rangle .$$

If the 2-qubit state $|\Phi_2\rangle$ is separable, it can be written as

$$|\Phi_2\rangle = |\psi_A\rangle \otimes |\psi_B\rangle ,$$

namely

$$c_{00}|01\rangle + c_{10}|10\rangle = \alpha_A\alpha_B|00\rangle + \alpha_A\beta_B|01\rangle + \beta_A\alpha_B|10\rangle + \beta_A\beta_B|11\rangle .$$

It follows

$$\alpha_A\alpha_B = \beta_A\beta_B = 0 , \qquad \alpha_A\beta_B = c_{01} , \qquad \beta_A\alpha_B = c_{10} .$$

These conditions are all satisfied only if $c_{01} = 0$ or $c_{10} = 0$, or both.

Further Reading

For special relativity:
W. Rindler, Introduction to Special Relativity (Oxford Univ. Press, Oxford, 1991).
For quantum mechanics:
C. Cohen-Tannoudji, B. Dui, and F. Laloe, Quantum Mechanics (Wiley, New York, 1991).
P.A.M. Dirac, The Principles of Quantum Mechanics (Oxford University Press, Oxford, 1982).
For quantum information:
M. Le Bellac, A Short Introduction to Quantum Information and Quantum Computation (Cambridge Univ. Press, Cambridge, 2006).
For quantum statistical mechanics:
K. Huang, Statistical Mechanics (Wiley, New York, 1987).

Chapter 2
Second Quantization of Light

In this chapter we discuss the quantization of electromagnetic waves, which we also denoted as light (visible or invisible to human eyes). After reviewing classical and quantum properties of the light in the vacuum, we discuss the so-called second quantization of the light field showing that this electromagnetic field can be expressed as a infinite sum of harmonic oscillators. These oscillators, which describe the possible frequencies of the radiation field, are quantized by introducing creation and annihilation operators acting on the Fock space of number representation. We analyze the Fock states of the radiation field and compare them with the coherent states. Finally, we consider two enlightening applications: the Casimir effect and the radiation field at finite temperature.

2.1 Electromagnetic Waves

The light is an electromagnetic field characterized by the coexisting presence of an electric field $\mathbf{E}(\mathbf{r}, t)$ and a magnetic field $\mathbf{B}(\mathbf{r}, t)$. From the equations of James Clerk Maxwell in vacuum and in the absence of sources, given by

$$\nabla \cdot \mathbf{E} = 0 , \tag{2.1}$$

$$\nabla \cdot \mathbf{B} = 0 , \tag{2.2}$$

$$\nabla \wedge \mathbf{E} = -\frac{\partial \mathbf{B}}{\partial t} , \tag{2.3}$$

$$\nabla \wedge \mathbf{B} = \varepsilon_0 \mu_0 \frac{\partial \mathbf{E}}{\partial t} , \tag{2.4}$$

© Springer International Publishing AG 2017
L. Salasnich, *Quantum Physics of Light and Matter*, UNITEXT for Physics,
DOI 10.1007/978-3-319-52998-1_2

one finds that the coupled electric and magnetic fields satisfy the wave equations

$$\left(\frac{1}{c^2}\frac{\partial^2}{\partial t^2} - \nabla^2\right)\mathbf{E} = \mathbf{0} \ , \tag{2.5}$$

$$\left(\frac{1}{c^2}\frac{\partial^2}{\partial t^2} - \nabla^2\right)\mathbf{B} = \mathbf{0} \ , \tag{2.6}$$

where

$$c = \frac{1}{\sqrt{\varepsilon_0\mu_0}} = 3 \times 10^8 \ \text{m/s} \tag{2.7}$$

is the speed of light in the vacuum. Note that the dielectric constant (electric permittivity) ε_0 and the magnetic constant (magnetic permeability) μ_0 are respectively $\varepsilon_0 = 8.85 \times 10^{-12} \, \text{C}^2/(\text{N m}^2)$ and $\mu_0 = 4\pi \times 10^{-7} \, \text{V s}/(\text{A m})$. Equations (2.5) and (2.6), which are fully confirmed by experiments, admit monochromatic complex plane wave solutions

$$\mathbf{E}(\mathbf{r}, t) = \mathbf{E}_0 \, e^{i(\mathbf{k}\cdot\mathbf{r}-\omega t)} \ , \tag{2.8}$$

$$\mathbf{B}(\mathbf{r}, t) = \mathbf{B}_0 \, e^{i(\mathbf{k}\cdot\mathbf{r}-\omega t)} \ , \tag{2.9}$$

where $\mathbf{k}$ is the wavevector and ω the angular frequency, such that

$$\omega = c k \ , \tag{2.10}$$

is the dispersion relation, with $k = |\mathbf{k}|$ is the wavenumber. From Maxwell's equations one finds that the vectors $\mathbf{E}$ and $\mathbf{B}$ are mutually orthogonal and such that

$$E = cB \ , \tag{2.11}$$

where $E = |\mathbf{E}|$ and $B = |\mathbf{B}|$. In addition they are transverse fields, i.e. orthogonal to the wavevector $\mathbf{k}$, which gives the direction of propagation of the wave. Notice that Eq. (2.11) holds for monochromatic plane waves but not in general. For completeness, let us remind that the wavelength λ is given by

$$\lambda = \frac{2\pi}{k} \ , \tag{2.12}$$

and that the linear frequency ν and the angular frequency $\omega = 2\pi\nu$ are related to the wavelength λ and to the wavenumber k by the formulas (Fig. 2.1)

$$\lambda \nu = \frac{\omega}{k} = c \ . \tag{2.13}$$

Fig. 2.1 Plot of a sinusoidal electromagnetic wave moving along the x axis

2.1.1 Photons

At the beginning of quantum mechanics Satyendra Nath Bose and Albert Einstein suggested that the light can be described as a gas of photons. A single photon of a monochromatic wave has the energy

$$\epsilon = h\nu = \hbar\omega ,\tag{2.14}$$

where $h = 6.63 \times 10^{-34}\,\text{J s}$ is the Planck constant and $\hbar = h/(2\pi) = 1.05 \times 10^{-34}\,\text{J s}$ is the reduced Planck constant. The linear momentum of the photon is given by the de Broglie relations

$$\mathbf{p} = \frac{h}{\lambda}\mathbf{n} = \hbar\mathbf{k} ,\tag{2.15}$$

where $\mathbf{n}$ is a unit vector in the direction of $\mathbf{k}$. Clearly, the energy of the photon can be written also as

$$\epsilon = pc ,\tag{2.16}$$

which is the energy one obtains for a relativistic particle of energy

$$\epsilon = \sqrt{m^2 c^4 + p^2 c^2} ,\tag{2.17}$$

setting to zero the rest mass, i.e. $m = 0$. The total energy H of monochromatic wave is given by

$$H = \sum_s \hbar\omega\, n_s ,\tag{2.18}$$

where n_s is the number of photons with angular frequency ω and polarization s in the monochromatic electromagnetic wave. Note that in general there are two possible polarizations: $s = 1, 2$, corresponding to two linearly independent orthogonal unit vectors ε_1 and ε_2 in the plane perpendicular to the wavevector $\mathbf{k}$.

A generic electromagnetic field is the superposition of many monochromatic electromagnetic waves. Calling ω_k the angular frequency of the monochromatic wave with wavenumber $\mathbf{k}$, the total energy H of the electromagnetic field is

$$H = \sum_k \sum_s \hbar\omega_k\, n_{\mathbf{k}s} ,\tag{2.19}$$

where $n_{\mathbf{k}s}$ is the number of photons with wavevector $\mathbf{k}$ and polarization s. The results derived here for the electromagnetic field, in particular Eq. (2.19), are called semi-classical because they do not take into account the so-called "quantum fluctuations of vacuum", i.e. the following remarkable experimental fact: photons can emerge from the vacuum of the electromagnetic field. To justify this property of the electromagnetic field one must perform the so-called second quantization of the field.

2.1.2 Electromagnetic Potentials and Coulomb Gauge

In full generality the electric field $\mathbf{E}(\mathbf{r}, t)$ and the magnetic field $\mathbf{B}(\mathbf{r}, t)$ can be expressed in terms of a scalar potential $\phi(\mathbf{r}, t)$ and a vector potential $\mathbf{A}(\mathbf{r}, t)$ as follows

$$\mathbf{E} = -\nabla\phi - \frac{\partial \mathbf{A}}{\partial t} , \tag{2.20}$$

$$\mathbf{B} = \nabla \wedge \mathbf{A} . \tag{2.21}$$

Actually these equations do not determine the electromagnetic potentials uniquely, since for an arbitrary scalar function $\Lambda(\mathbf{r}, t)$ the so-called "gauge transformation"

$$\phi \rightarrow \phi' = \phi + \frac{\partial \Lambda}{\partial t} , \tag{2.22}$$

$$\mathbf{A} \rightarrow \mathbf{A}' = \mathbf{A} - \nabla\Lambda , \tag{2.23}$$

leaves the fields $\mathbf{E}$ and $\mathbf{B}$ unaltered. There is thus an infinite number of different electromagnetic potentials that correspond to a given configuration of measurable fields. We use this remarkable property to choose a gauge transformation such that

$$\nabla \cdot \mathbf{A} = 0 . \tag{2.24}$$

This condition defines the Coulomb (or radiation) gauge, and the vector field $\mathbf{A}$ is called transverse field. For a complex monochromatic plane wave

$$\mathbf{A}(\mathbf{r}, t) = \mathbf{A}_0 \, e^{i(\mathbf{k}\cdot\mathbf{r}-\omega t)} \tag{2.25}$$

the Coulomb gauge (2.24) gives
$$\mathbf{k} \cdot \mathbf{A} = 0 , \tag{2.26}$$

i.e. $\mathbf{A}$ is perpendicular (transverse) to the wavevector $\mathbf{k}$. In the vacuum and without sources, from the first Maxwell equation (2.1) and Eq. (2.20) one immediately finds

$$\nabla^2\phi + \frac{\partial}{\partial t} (\nabla \cdot \mathbf{A}) = 0 , \tag{2.27}$$

and under the Coulomb gauge (2.24) one gets

$$\nabla^2 \phi = 0 . \tag{2.28}$$

Imposing that the scalar potential is zero at infinity, this Laplace's equation has the unique solution

$$\phi(\mathbf{r}, t) = 0 , \tag{2.29}$$

and consequently

$$\mathbf{E} = -\frac{\partial \mathbf{A}}{\partial t} , \tag{2.30}$$

$$\mathbf{B} = \nabla \wedge \mathbf{A} . \tag{2.31}$$

Thus, in the Coulomb gauge one needs only the electromagnetic vector potential $\mathbf{A}(\mathbf{r}, t)$ to obtain the electromagnetic field if there are no charges and no currents. Notice that here $\mathbf{E}$ and $\mathbf{B}$ are transverse fields like $\mathbf{A}$, which satisfy Eqs. (2.5) and (2.6). The electromagnetic field described by these equations is often called the radiation field, and also the vector potential satisfies the wave equation

$$\left(\frac{1}{c^2} \frac{\partial^2}{\partial t^2} - \nabla^2 \right) \mathbf{A} = \mathbf{0} . \tag{2.32}$$

We now expand the vector potential $\mathbf{A}(\mathbf{r}, t)$ as a Fourier series of monochromatic plane waves. The vector potential is a real vector field, i.e. $\mathbf{A} = \mathbf{A}^*$ and consequently we write

$$\mathbf{A}(\mathbf{r}, t) = \sum_{\mathbf{k}} \sum_{s} \left[A_{\mathbf{k}s}(t) \frac{e^{i\mathbf{k}\cdot\mathbf{r}}}{\sqrt{V}} + A_{\mathbf{k}s}^*(t) \frac{e^{-i\mathbf{k}\cdot\mathbf{r}}}{\sqrt{V}} \right] \varepsilon_{\mathbf{k}s} , \tag{2.33}$$

where $A_{\mathbf{k}s}(t)$ and $A_{\mathbf{k}s}^*(t)$ are the dimensional complex conjugate coefficients of the expansion, the complex plane waves $e^{i\mathbf{k}\cdot\mathbf{r}}/\sqrt{V}$ normalized in a volume V are the basis functions of the expansion, and $\varepsilon_{\mathbf{k}1}$ and $\varepsilon_{\mathbf{k}2}$ are two mutually orthogonal real unit vectors of polarization which are also orthogonal to $\mathbf{k}$ (transverse polarization vectors).

Taking into account Eqs. (2.30) and (2.31) we get

$$\mathbf{E}(\mathbf{r}, t) = -\sum_{\mathbf{k}} \sum_{s} \left[\dot{A}_{\mathbf{k}s}(t) \frac{e^{i\mathbf{k}\cdot\mathbf{r}}}{\sqrt{V}} + \dot{A}_{\mathbf{k}s}^*(t) \frac{e^{-i\mathbf{k}\cdot\mathbf{r}}}{\sqrt{V}} \right] \varepsilon_{\mathbf{k}s} , \tag{2.34}$$

$$\mathbf{B}(\mathbf{r}, t) = \sum_{\mathbf{k}} \sum_{s} \left[A_{\mathbf{k}s}(t) \frac{e^{i\mathbf{k}\cdot\mathbf{r}}}{\sqrt{V}} - A_{\mathbf{k}s}^*(t) \frac{e^{-i\mathbf{k}\cdot\mathbf{r}}}{\sqrt{V}} \right] i\mathbf{k} \wedge \varepsilon_{\mathbf{k}s} , \tag{2.35}$$

where both vector fields are explicitly real fields. Moreover, inserting Eq. (2.33) into Eq. (2.32) we recover familiar differential equations of decoupled harmonic oscillators:

$$\ddot{A}_{ks}(t) + \omega_k^2 A_{ks}(t) = 0 , \tag{2.36}$$

with $\omega_k = ck$, which have the complex solutions

$$A_{ks}(t) = A_{ks}(0) e^{-i\omega_k t} . \tag{2.37}$$

These are the complex amplitudes of the infinite harmonic normal modes of the radiation field.

A familiar result of electromagnetism is that the classical energy of the electromagnetic field in vacuum is given by

$$H = \int d^3 \mathbf{r} \left(\frac{\varepsilon_0}{2} \mathbf{E}(\mathbf{r}, t)^2 + \frac{1}{2\mu_0} \mathbf{B}(\mathbf{r}, t)^2 \right) , \tag{2.38}$$

namely

$$H = \int d^3 \mathbf{r} \left(\frac{\varepsilon_0}{2} \left(\frac{\partial \mathbf{A}(\mathbf{r}, t)}{\partial t} \right)^2 + \frac{1}{2\mu_0} (\nabla \wedge \mathbf{A}(\mathbf{r}, t))^2 \right) \tag{2.39}$$

by using the Maxwell equations in the Coulomb gauge (2.30) and (2.31). Inserting into this expression Eq. (2.33) or Eqs. (2.34) and (2.35) into Eq. (2.38) we find

$$H = \sum_{\mathbf{k}} \sum_{s} \varepsilon_0 \omega_k^2 \left(A_{ks}^* A_{ks} + A_{ks} A_{ks}^* \right) . \tag{2.40}$$

It is now convenient to introduce dimensionless complex coefficients $a_{ks}(t)$ and $a_{ks}^*(t)$ related to the dimensional complex coefficients $A_{ks}(t)$ and $A_{ks}^*(t)$ by

$$A_{ks}(t) = \sqrt{\frac{\hbar}{2\varepsilon_0 \omega_k}} a_{ks}(t) . \tag{2.41}$$

$$A_{ks}^*(t) = \sqrt{\frac{\hbar}{2\varepsilon_0 \omega_k}} a_{ks}^*(t) . \tag{2.42}$$

In this way the energy H reads

$$H = \sum_{\mathbf{k}} \sum_{s} \frac{\hbar \omega_k}{2} \left(a_{ks}^* a_{ks} + a_{ks} a_{ks}^* \right) . \tag{2.43}$$

This energy is actually independent on time: the time dependence of the complex amplitudes $a_{ks}^*(t)$ and $a_{ks}(t)$ cancels due to Eq. (2.37). Instead of using the complex amplitudes $a_{ks}^*(t)$ and $a_{ks}(t)$ one can introduce the real variables

$$q_{\mathbf{k}s}(t) = \sqrt{\frac{2\hbar}{\omega_k}} \frac{1}{2} \left(a_{\mathbf{k}s}(t) + a_{\mathbf{k}s}^*(t) \right) \tag{2.44}$$

$$p_{\mathbf{k}s}(t) = \sqrt{2\hbar\omega_k} \frac{1}{2i} \left(a_{\mathbf{k}s}(t) - a_{\mathbf{k}s}^*(t) \right) \tag{2.45}$$

such that the energy of the radiation field reads

$$H = \sum_{\mathbf{k}} \sum_{s} \left(\frac{p_{\mathbf{k},s}^2}{2} + \frac{1}{2}\omega_k^2 q_{\mathbf{k}s}^2 \right). \tag{2.46}$$

This energy resembles that of infinitely many harmonic oscillators with unitary mass and frequency ω_k. It is written in terms of an infinite set of real harmonic oscillators: two oscillators (due to polarization) for each mode of wavevector $\mathbf{k}$ and angular frequency ω_k.

2.2 Second Quantization of Light

In 1927 Paul Dirac performed the quantization of the classical Hamiltonian (2.46) by promoting the real coordinates $q_{\mathbf{k}s}$ and the real momenta $p_{\mathbf{k}s}$ to operators:

$$q_{\mathbf{k}s} \rightarrow \hat{q}_{\mathbf{k}s}, \tag{2.47}$$

$$p_{\mathbf{k}s} \rightarrow \hat{p}_{\mathbf{k}s}, \tag{2.48}$$

satisfying the commutation relations

$$[\hat{q}_{\mathbf{k}s}, \hat{p}_{\mathbf{k}'s'}] = i\hbar \, \delta_{\mathbf{k},\mathbf{k}'} \, \delta_{s,s'}, \tag{2.49}$$

where $[\hat{A}, \hat{B}] = \hat{A}\hat{B} - \hat{B}\hat{A}$. The quantum Hamiltonian is thus given by

$$\hat{H} = \sum_{\mathbf{k}} \sum_{s} \left(\frac{\hat{p}_{\mathbf{k},s}^2}{2} + \frac{1}{2}\omega_k^2 \hat{q}_{\mathbf{k}s}^2 \right). \tag{2.50}$$

Following a standard approach for the canonical quantization of the Harmonic oscillator, we introduce annihilation and creation operators

$$\hat{a}_{\mathbf{k}s} = \sqrt{\frac{\omega_k}{2\hbar}} \left(\hat{q}_{\mathbf{k}s} + \frac{i}{\omega_k} \hat{p}_{\mathbf{k}s} \right), \tag{2.51}$$

$$\hat{a}_{\mathbf{k}s}^+ = \sqrt{\frac{\omega_k}{2\hbar}} \left(\hat{q}_{\mathbf{k}s} - \frac{i}{\omega_k} \hat{p}_{\mathbf{k}s} \right), \tag{2.52}$$

which satisfy the commutation relations

$$[\hat{a}_{\mathbf{k}s}, \hat{a}^+_{\mathbf{k}'s'}] = \delta_{\mathbf{k},\mathbf{k}'}\,\delta_{s,s'}\,, \tag{2.53}$$

and the quantum Hamiltonian (2.50) becomes

$$\hat{H} = \sum_{\mathbf{k}}\sum_{s} \hbar\omega_k \left(\hat{a}^+_{\mathbf{k}s}\hat{a}_{\mathbf{k}s} + \frac{1}{2}\right)\,. \tag{2.54}$$

Obviously this quantum Hamiltonian can be directly obtained from the classical one, given by Eq. (2.43), by promoting the complex amplitudes $a_{\mathbf{k}s}$ and $a^*_{\mathbf{k}s}$ to operators:

$$a_{\mathbf{k}s} \to \hat{a}_{\mathbf{k}s}\,, \tag{2.55}$$

$$a^*_{\mathbf{k}s} \to \hat{a}^+_{\mathbf{k}s}\,, \tag{2.56}$$

satisfying the commutation relations (2.53).

The operators $\hat{a}_{\mathbf{k}s}$ and $\hat{a}^+_{\mathbf{k}s}$ act in the Fock space $\mathcal{F}$, i.e. the infinite dimensional Hilbert space of "number representation" introduced in 1932 by Vladimir Fock. A generic state of this Fock space $\mathcal{F}$ is given by

$$| \ldots n_{\mathbf{k}s} \ldots n_{\mathbf{k}'s'} \ldots n_{\mathbf{k}''s''} \ldots \rangle\,, \tag{2.57}$$

meaning that there are $n_{\mathbf{k}s}$ photons with wavevector $\mathbf{k}$ and polarization s, $n_{\mathbf{k}'s'}$ photons with wavevector $\mathbf{k}'$ and polarization s', $n_{\mathbf{k}''s''}$ photons with wavevector $\mathbf{k}''$ and polarization s'', et cetera. The Fock space $\mathcal{F}$ is given by

$$\mathcal{F} = \mathcal{H}_0 \oplus \mathcal{H}_1 \oplus \mathcal{H}_2 \oplus \mathcal{H}_3 \oplus \cdots \oplus \mathcal{H}_\infty = \bigoplus_{n=0}^{\infty} \mathcal{H}_n\,, \tag{2.58}$$

where

$$\mathcal{H}_n = \mathcal{H} \otimes \mathcal{H} \otimes \cdots \otimes \mathcal{H} = \mathcal{H}^{\otimes n} \tag{2.59}$$

is the Hilbert space of n identical photons, which is n times the tensor product $\otimes$ of the single-photon Hilbert space $\mathcal{H} = \mathcal{H}^{\otimes 1}$. Thus, $\mathcal{F}$ is the infinite direct sum $\oplus$ of increasing n-photon Hilbert states $\mathcal{H}_n$, and we can formally write

$$\mathcal{F} = \bigoplus_{n=0}^{\infty} \mathcal{H}^{\otimes n}\,. \tag{2.60}$$

Notice that in the definition of the Fock space $\mathcal{F}$ one must include the space $\mathcal{H}_0 = \mathcal{H}^{\otimes 0}$, which is the Hilbert space of 0 photons, containing only the vacuum state

$$|0\rangle = | \ldots 0 \ldots 0 \ldots 0 \ldots \rangle\,, \tag{2.61}$$

and its dilatations $\gamma|0\rangle$ with γ a generic complex number. The operators $\hat{a}_{\mathbf{k}s}$ and $\hat{a}_{\mathbf{k}s}^+$ are called annihilation and creation operators because they respectively destroy and create one photon with wavevector $\mathbf{k}$ and polarization s, namely

$$\hat{a}_{\mathbf{k}s}|\ldots n_{\mathbf{k}s}\ldots\rangle = \sqrt{n_{\mathbf{k}s}}|\ldots n_{\mathbf{k}s}-1\ldots\rangle\,, \tag{2.62}$$

$$\hat{a}_{\mathbf{k}s}^+|\ldots n_{\mathbf{k}s}\ldots\rangle = \sqrt{n_{\mathbf{k}s}+1}|\ldots n_{\mathbf{k}s}+1\ldots\rangle\,. \tag{2.63}$$

Note that these properties follow directly from the commutation relations (2.53). Consequently, for the vacuum state $|0\rangle$ one finds

$$\hat{a}_{\mathbf{k}s}|0\rangle = 0_F\,, \tag{2.64}$$

$$\hat{a}_{\mathbf{k}s}^+|0\rangle = |1_{\mathbf{k}s}\rangle = |\mathbf{k}s\rangle\,, \tag{2.65}$$

where 0_F is the zero of the Fock space (usually indicated with 0), and $|\mathbf{k}s\rangle$ is clearly the state of one photon with wavevector $\mathbf{k}$ and polarization s, such that

$$\langle\mathbf{r}|\mathbf{k}s\rangle = \frac{e^{i\mathbf{k}\cdot\mathbf{r}}}{\sqrt{V}}\varepsilon_{\mathbf{k}s}\,. \tag{2.66}$$

From Eqs. (2.62) and (2.63) it follows immediately that

$$\hat{N}_{\mathbf{k}s} = \hat{a}_{\mathbf{k}s}^+\hat{a}_{\mathbf{k}s} \tag{2.67}$$

is the number operator which counts the number of photons in the single-particle state $|\mathbf{k}s\rangle$, i.e.

$$\hat{N}_{\mathbf{k}s}|\ldots n_{\mathbf{k}s}\ldots\rangle = n_{\mathbf{k}s}|\ldots n_{\mathbf{k}s}\ldots\rangle\,. \tag{2.68}$$

Notice that the quantum Hamiltonian of the light can be written as

$$\hat{H} = \sum_{\mathbf{k}}\sum_{s}\hbar\omega_k\left(\hat{N}_{\mathbf{k}s}+\frac{1}{2}\right)\,, \tag{2.69}$$

and this expression is very similar, but not equal, to the semiclassical formula (2.19). The differences are that $\hat{N}_{\mathbf{k}s}$ is a quantum number operator and that the energy E_{vac} of the vacuum state $|0\rangle$ is not zero but is instead given by

$$E_{vac} = \sum_{\mathbf{k}}\sum_{s}\frac{1}{2}\hbar\omega_k\,. \tag{2.70}$$

A quantum harmonic oscillator of frequency $\omega_{\mathbf{k}}$ has a finite minimal energy $\hbar\omega_{\mathbf{k}}$, which is called zero-point energy. In the case of the quantum electromagnetic field there is an infinite number of harmonic oscillators and the total zero-point energy, given by Eq. (2.70), is clearly infinite. The infinite constant E_{vac} is usually eliminated by simply shifting to zero the energy associated to the vacuum state $|0\rangle$.

The quantum electric and magnetic fields can be obtained from the classical expressions, Eqs. (2.34) and (2.35), taking into account the quantization of the classical complex amplitudes a_{ks} and a_{ks}^* and their time-dependence, given by Eq. (2.37) and its complex conjugate. In this way we obtain

$$\hat{\mathbf{E}}(\mathbf{r}, t) = i \sum_{\mathbf{k}} \sum_{s} \sqrt{\frac{\hbar \omega_k}{2\varepsilon_0 V}} \left[\hat{a}_{\mathbf{k}s} \, e^{i(\mathbf{k}\cdot\mathbf{r}-\omega_k t)} - \hat{a}_{\mathbf{k}s}^+ \, e^{-(i\mathbf{k}\cdot\mathbf{r}-\omega_k t)} \right] \varepsilon_{\mathbf{k}s} , \qquad (2.71)$$

$$\hat{\mathbf{B}}(\mathbf{r}, t) = \sum_{\mathbf{k}} \sum_{s} \sqrt{\frac{\hbar}{2\varepsilon_0 \omega_k V}} \left[\hat{a}_{\mathbf{k}s} \, e^{i(\mathbf{k}\cdot\mathbf{r}-\omega_k t)} - \hat{a}_{\mathbf{k}s}^+ \, e^{-i(\mathbf{k}\cdot\mathbf{r}-\omega_k t)} \right] i \frac{\mathbf{k}}{|\mathbf{k}|} \wedge \varepsilon_{\mathbf{k}s} .$$

$$(2.72)$$

It is important to stress that the results of our canonical quantization of the radiation field suggest a remarkable philosophical idea: there is a unique quantum electromagnetic field in the universe and all the photons we see are the massless particles associated to it. In fact, the quantization of the electromagnetic field is the first step towards the so-called quantum field theory or second quantization of fields, where all particles in the universe are associated to few quantum fields and their corresponding creation and annihilation operators.

2.2.1 Fock versus Coherent States for the Light Field

Let us now consider for simplicity a linearly polarized monochromatic wave of the radiation field. For instance, let us suppose that the direction of polarization is given by the vector ε. From Eq. (2.71) one finds immediately that the quantum electric field can be then written in a simplified notation as

$$\hat{\mathbf{E}}(\mathbf{r}, t) = \sqrt{\frac{\hbar \omega}{2\varepsilon_0 V}} i \left[\hat{a} \, e^{i(\mathbf{k}\cdot\mathbf{r}-\omega t)} - \hat{a}^+ \, e^{-i(\mathbf{k}\cdot\mathbf{r}-\omega t)} \right] \varepsilon \qquad (2.73)$$

where $\omega = \omega_k = c|\mathbf{k}|$. Notice that, to simplify the notation, we have removed the subscripts in the annihilation and creation operators $\hat{a}$ and $\hat{a}^+$. If there are exactly n photons in this polarized monochromatic wave the Fock state of the system is given by

$$|n\rangle = \frac{1}{\sqrt{n!}} \left(\hat{a}^+ \right)^n |0\rangle . \qquad (2.74)$$

It is then straightforward to show, by using Eqs. (2.62) and (2.63), that

$$\langle n | \hat{\mathbf{E}}(\mathbf{r}, t) | n \rangle = \mathbf{0}, \qquad (2.75)$$

for all values of the photon number n, no matter how large. This result holds for all modes, which means then that the expectation value of the electric field in any many-photon Fock state is zero. On the other hand, the expectation value of $\hat{E}(\mathbf{r}, t)^2$ is given by

$$\langle n|\hat{E}(\mathbf{r}, t)^2|n\rangle = \frac{\hbar\omega}{\varepsilon_0 V}\left(n + \frac{1}{2}\right) . \tag{2.76}$$

For this case, as for the energy, the expectation value is nonvanishing even when $n = 0$, with the result that for the total field (2.71), consisting of an infinite number of modes, the expectation value of $\hat{E}(\mathbf{r}, t)^2$ is infinite for all many-photon states, including the vacuum state. Obviously a similar reasoning applies for the magnetic field (2.72) and, as discussed in the previous section, the zero-point constant is usually removed.

One must remember that the somehow strange result of Eq. (2.75) is due to the fact that the expectation value is performed with the Fock state $|n\rangle$, which means that the number of photons is fixed because

$$\hat{N}|n\rangle = n|n\rangle . \tag{2.77}$$

Nevertheless, usually the number of photons in the radiation field is not fixed, in other words the system is not in a pure Fock state. For example, the radiation field of a well-stabilized laser device operating in a single mode is described by a coherent state $|\alpha\rangle$, such that

$$\hat{a}|\alpha\rangle = \alpha|\alpha\rangle , \tag{2.78}$$

with

$$\langle\alpha|\alpha\rangle = 1 . \tag{2.79}$$

The coherent state $|\alpha\rangle$, introduced in 1963 by Roy Glauber, is thus the eigenstate of the annihilation operator $\hat{a}$ with complex eigenvalue $\alpha = |\alpha|e^{i\theta}$. $|\alpha\rangle$ does not have a fixed number of photons, i.e. it is not an eigenstate of the number operator $\hat{N}$, and it is not difficult to show that $|\alpha\rangle$ can be expanded in terms of number (Fock) states $|n\rangle$ as follows

$$|\alpha\rangle = e^{-|\alpha|^2/2} \sum_{n=0}^{\infty} \frac{\alpha^n}{\sqrt{n!}}|n\rangle . \tag{2.80}$$

From Eq. (2.78) one immediately finds

$$\bar{N} = \langle\alpha|\hat{N}|\alpha\rangle = |\alpha|^2 , \tag{2.81}$$

and it is natural to set

$$\alpha = \sqrt{\bar{N}}\, e^{i\theta} , \tag{2.82}$$

where $\bar{N}$ is the average number of photons in the coherent state, while θ is the phase of the coherent state. For the sake of completeness, we observe that

$$\langle \alpha | \hat{N}^2 | \alpha \rangle = |\alpha|^2 + |\alpha|^4 = \bar{N} + \bar{N}^2 \tag{2.83}$$

and consequently

$$\langle \alpha | \hat{N}^2 | \alpha \rangle - \langle \alpha | \hat{N} | \alpha \rangle^2 = \bar{N} \; , \tag{2.84}$$

while

$$\langle n | \hat{N}^2 | n \rangle = n^2 \tag{2.85}$$

and consequently

$$\langle n | \hat{N}^2 | n \rangle - \langle n | \hat{N} | n \rangle^2 = 0 \; . \tag{2.86}$$

The expectation value of the electric field $\hat{\mathbf{E}}(\mathbf{r}, t)$ of the linearly polarized monochromatic wave, Eq. (2.73), in the coherent state $|\alpha\rangle$ reads

$$\langle \alpha | \hat{\mathbf{E}}(\mathbf{r}, t) | \alpha \rangle = -\sqrt{\frac{2\bar{N}\hbar\omega}{\varepsilon_0 V}} \, \sin(\mathbf{k} \cdot \mathbf{r} - \omega t + \theta) \, \boldsymbol{\varepsilon} \; , \tag{2.87}$$

while the expectation value of $\hat{E}(\mathbf{r}, t)^2$ is given by

$$\langle \alpha | \hat{E}(\mathbf{r}, t)^2 | \alpha \rangle = \frac{2\bar{N}\hbar\omega}{\varepsilon_0 V} \, \sin^2(\mathbf{k} \cdot \mathbf{r} - \omega t + \theta) \; . \tag{2.88}$$

These results suggest that the coherent state is indeed a useful tool to investigate the correspondence between quantum field theory and classical field theory. Indeed, in 1965 at the University of Milan Fortunato Tito Arecchi experimentally verified that a single-mode laser is in a coherent state with a definite but unknown phase. Thus, the coherent state gives photocount statistics that are in accord with laser experiments and has coherence properties similar to those of a classical field, which are useful for explaining interference effects.

To conclude this subsection we observe that we started the quantization of the light field by expanding the vector potential $\mathbf{A}(\mathbf{r}, t)$ as a Fourier series of monochromatic plane waves. But this is not the unique choice. Indeed, one can expand the vector potential $\mathbf{A}(\mathbf{r}, t)$ by using any orthonormal set of wavefunctions (spatio-temporal modes) which satisfy the Maxwell equations. After quantization, the expansion coefficients of the single mode play the role of creation and annihilation operators of that mode.

2.2.2 Linear and Angular Momentum of the Radiation Field

In addition to the total energy there are other interesting conserved quantities which characterize the classical radiation field. They are the total linear momentum

$$\mathbf{P} = \int d^3\mathbf{r}\, \varepsilon_0\, \mathbf{E}(\mathbf{r}, t) \wedge \mathbf{B}(\mathbf{r}, t) \tag{2.89}$$

and the total angular momentum

$$\mathbf{J} = \int d^3\mathbf{r}\, \varepsilon_0\, \mathbf{r} \wedge (\mathbf{E}(\mathbf{r}, t) \wedge \mathbf{B}(\mathbf{r}, t)) \ . \tag{2.90}$$

By using the canonical quantization one easily finds that the quantum total linear momentum operator is given by

$$\hat{\mathbf{P}} = \sum_{\mathbf{k}} \sum_{s} \hbar\mathbf{k} \left(\hat{N}_{\mathbf{k}s} + \frac{1}{2} \right) , \tag{2.91}$$

where $\hbar\mathbf{k}$ is the linear momentum of a photon of wavevector $\mathbf{k}$.

The quantization of the total angular momentum $\mathbf{J}$ is a more intricate problem. In fact, the linearly polarized states $|\mathbf{k}s\rangle$ (with $s = 1, 2$) are eigenstates of $\hat{H}$ and $\hat{\mathbf{P}}$, but they are not eigenstates of $\hat{\mathbf{J}}$, nor of $\hat{J}^2$, and nor of $\hat{J}_z$. The eigenstates of $\hat{J}^2$ and $\hat{J}_z$ do not have a fixed wavevector $\mathbf{k}$ but only a fixed wavenumber $k = |\mathbf{k}|$. These states, which can be indicated as $|kjm_js\rangle$, are associated to vector spherical harmonics $\mathbf{Y}_{j,m_j}(\theta, \phi)$, with θ and ϕ the spherical angles of the wavevector $\mathbf{k}$. These states $|kjm_js\rangle$ are not eigenstates of $\hat{\mathbf{P}}$ but they are eigenstates of $\hat{H}$, $\hat{J}^2$ and $\hat{J}_z$, namely

$$\hat{H}|kjm_js\rangle = ck|kjm_js\rangle \tag{2.92}$$

$$\hat{J}^2|kjm_j\rangle = j(j+1)\hbar^2|kjm_js\rangle \tag{2.93}$$

$$\hat{J}_z|kjm_js\rangle = m_j\hbar|kjm_js\rangle \tag{2.94}$$

with $m_j = -j, -j+1, \ldots, j-1, j$ and $j = 1, 2, 3, \ldots$. In problems where the charge distribution which emits the electromagnetic radiation has spherical symmetry it is indeed more useful to expand the radiation field in vector spherical harmonics than in plane waves. Clearly, one can write

$$|\mathbf{k}s\rangle = \sum_{jm_j} c_{jm_j}(\theta, \phi)|kjm_js\rangle , \tag{2.95}$$

where the coefficients $c_{jm_j}(\theta, \phi)$ of this expansion give the amplitude probability of finding a photon of fixed wavenumber $k = |\mathbf{k}|$ and orbital quantum numbers j and m_j with spherical angles θ and ϕ.

2.2.3 Zero-Point Energy and the Casimir Effect

There are situations in which the zero-point energy E_{vac} of the electromagnetic field oscillators give rise to a remarkable quantum phenomenon: the Casimir effect.

The zero-point energy (2.70) of the electromagnetic field in a region of volume V can be written as

$$E_{vac} = \sum_{\mathbf{k}} \sum_{s} \frac{1}{2} \hbar c \sqrt{k_x^2 + k_y^2 + k_z^2} = V \int \frac{d^3\mathbf{k}}{(2\pi)^3} \hbar c \sqrt{k_x^2 + k_y^2 + k_z^2} \qquad (2.96)$$

by using the dispersion relation $\omega_k = ck = c\sqrt{k_y^2 + k_y^2 + k_z^2}$. In particular, considering a region with the shape of a parallelepiped of length L along both x and y and length a along z, the volume V is given by $V = L^2 a$ and the vacuum energy E_{vac} in the region is

$$\begin{aligned}
E_{vac} &= \hbar c \int_{-\infty}^{+\infty} \frac{L\,dk_x}{2\pi} \int_{-\infty}^{+\infty} \frac{L\,dk_y}{2\pi} \int_{-\infty}^{+\infty} \frac{a\,dk_z}{2\pi} \sqrt{k_x^2 + k_y^2 + k_z^2} \\
&= \frac{\hbar c}{2\pi} L^2 \int_0^\infty dk_\| k_\| \left[\int_0^\infty dn \sqrt{k_\|^2 + \frac{n^2\pi^2}{a^2}} \right],
\end{aligned} \qquad (2.97)$$

where the second expression is obtained setting $k_\| = \sqrt{k_x^2 + k_y^2}$ and $n = (a/\pi)k_z$.

Let us now consider the presence of two perfect metallic plates with the shape of a square of length L having parallel faces lying in the (x, y) plane at distance a. Along the z axis the stationary standing waves of the electromagnetic field vanishes on the metal plates and the k_z component of the wavevector $\mathbf{k}$ is no more a continuum variabile but it is quantized via

$$k_z = n\frac{\pi}{a}, \qquad (2.98)$$

where now $n = 0, 1, 2, \ldots$ is an integer number, and not a real number as in Eq. (2.97). In this case the zero-point energy in the volume $V = L^2 a$ between the two plates reads

$$E'_{vac} = \frac{\hbar c}{2\pi} L^2 \int_0^\infty dk_\| k_\| \left[\frac{k_\|}{2} + \sum_{n=1}^\infty \sqrt{k_\|^2 + \frac{n^2\pi^2}{a^2}} \right], \qquad (2.99)$$

where for $k_z = 0$ the state $|(k_z, k_y, 0)1\rangle$ with polarization $\epsilon_{\mathbf{k}1}$ parallel to the z axis (and orthogonal to the metallic plates) is bound, while the state $|(k_z, k_y, 0)2\rangle$ with polarization $\epsilon_{\mathbf{k}2}$ orthogonal to the z axis (and parallel to the metallic plates) is not bound and it does not contribute to the discrete summation (Fig. 2.2).

Fig. 2.2 Graphical representation of the parallel plates in the Casimir effect

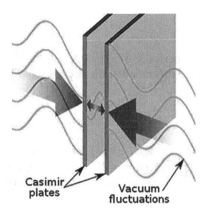

Casimir plates

Vacuum fluctuations

The difference between E'_{vac} and E_{vac} divided by L^2 gives the net energy per unit surface area $\mathcal{E}$, namely

$$\mathcal{E} = \frac{E'_{vac} - E_{vac}}{L^2} = \frac{\hbar c}{2\pi} \left(\frac{\pi}{a}\right)^3 \left[\frac{1}{2}A(0) + \sum_{n=1}^{\infty} A(n) - \int_0^{\infty} dn\, A(n)\right], \quad (2.100)$$

where we have defined

$$A(n) = \int_0^{+\infty} d\zeta\, \zeta \sqrt{\zeta^2 + n^2} = \frac{1}{3}\left[(n^2 + \infty)^{3/2} - n^2\right], \quad (2.101)$$

with $\zeta = (a/\pi)k_{\parallel}$. Notice that $A(n)$ is clearly divergent but Eq. (2.100) is not divergent due to the cancellation of divergences with opposite sign. In fact, by using the Euler-MacLaurin formula for the difference of infinite series and integrals

$$\frac{1}{2}A(0) + \sum_{n=1}^{\infty} A(n) - \int_0^{\infty} dn\, A(n) = -\frac{1}{6 \cdot 2!}\frac{dA}{dn}(0) + \frac{1}{30 \cdot 4!}\frac{d^3 A}{dn^3}(0) - \frac{1}{42 \cdot 6!}\frac{d^5 A}{dn^5}(0) + \cdots$$

$$(2.102)$$

and Eq. (2.101) from which

$$\frac{dA}{dn}(0) = 0, \qquad \frac{d^3 A}{dn^3}(0) = -2, \qquad \frac{d^5 A}{dn^5}(0) = 0 \qquad (2.103)$$

and all higher derivatives of $A(n)$ are zero, one eventually obtains

$$\mathcal{E} = -\frac{\pi^2}{720}\frac{\hbar c}{a^3}. \quad (2.104)$$

From this energy difference $\mathcal{E}$ one deduces that there is an attractive force per unit area $\mathcal{F}$ between the two plates, given by

$$\mathcal{F} = -\frac{d\mathcal{E}}{da} = -\frac{\pi^2}{240}\frac{\hbar c}{a^4} . \tag{2.105}$$

Numerically this result, predicted in 1948 by Hendrik Casimir during his research
activity at the Philips Physics Laboratory in Eindhoven, is very small

$$\mathcal{F} = -\frac{1.30 \times 10^{-27}\text{N m}^2}{a^4} . \tag{2.106}$$

Nevertheless, it has been experimentally verified by Steven Lamoreaux in 1997 at
the University of Washington and by Giacomo Bressi, Gianni Carugno, Roberto
Onofrio, and Giuseppe Ruoso in 2002 at the University of Padua.

2.3 Quantum Radiation Field at Finite Temperature

Let us consider the quantum radiation field in thermal equilibrium with a bath at the
temperature T. The relevant quantity to calculate all thermodynamical properties of
the system is the grand-canonical partition function $\mathcal{Z}$, given by

$$\mathcal{Z} = Tr[e^{-\beta(\hat{H}-\mu\hat{N})}] \tag{2.107}$$

where $\beta = 1/(k_B T)$ with $k_B = 1.38 \times 10^{-23}$ J/K the Boltzmann constant,

$$\hat{H} = \sum_k \sum_s \hbar\omega_k \hat{N}_{ks} , \tag{2.108}$$

is the quantum Hamiltonian without the zero-point energy,

$$\hat{N} = \sum_k \sum_s \hat{N}_{ks} \tag{2.109}$$

is the total number operator, and μ is the chemical potential, fixed by the conservation
of the particle number. For photons $\mu = 0$ and consequently the number of photons
is not fixed. This implies that

$$\mathcal{Z} = \sum_{\{n_{ks}\}}\langle \dots n_{ks} \dots |e^{-\beta\hat{H}}| \dots n_{ks} \dots\rangle = \sum_{\{n_{ks}\}}\langle \dots n_{ks} \dots |e^{-\beta\sum_{ks}\hbar\omega_k\hat{N}_{ks}}| \dots n_{ks} \dots\rangle$$

$$= \sum_{\{n_{ks}\}} e^{-\beta\sum_{ks}\hbar\omega_k n_{ks}} = \sum_{\{n_{ks}\}}\prod_{ks} e^{-\beta\hbar\omega_k n_{ks}} = \prod_{ks}\sum_{n_{ks}} e^{-\beta\hbar\omega_k n_{ks}} = \prod_{ks}\sum_{n=0}^{\infty} e^{-\beta\hbar\omega_k n}$$

$$= \prod_{ks}\frac{1}{1-e^{-\beta\hbar\omega_k}} . \tag{2.110}$$

Quantum statistical mechanics dictates that the thermal average of any operator $\hat{A}$ is obtained as

$$\langle \hat{A} \rangle_T = \frac{1}{\mathcal{Z}} Tr[\hat{A} \, e^{-\beta(\hat{H}-\mu\hat{N})}] \, . \tag{2.111}$$

In our case the calculations are simplified because $\mu = 0$. Let us suppose that $\hat{A} = \hat{H}$, it is then quite easy to show that

$$\langle \hat{H} \rangle_T = \frac{1}{\mathcal{Z}} Tr[\hat{H} \, e^{-\beta\hat{H}}] = -\frac{\partial}{\partial\beta} \ln\left(Tr[e^{-\beta\hat{H}}]\right) = -\frac{\partial}{\partial\beta} \ln(\mathcal{Z}) \, . \tag{2.112}$$

By using Eq. (2.110) we immediately obtain

$$\ln(\mathcal{Z}) = -\sum_k \sum_s \ln\left(1 - e^{-\beta\hbar\omega_k}\right) , \tag{2.113}$$

and finally from Eq. (2.112) we get

$$\langle \hat{H} \rangle_T = \sum_k \sum_s \frac{\hbar\omega_k}{e^{\beta\hbar\omega_k} - 1} = \sum_k \sum_s \hbar\omega_k \, \langle \hat{N}_{ks} \rangle_T \, . \tag{2.114}$$

In the continuum limit, where

$$\sum_k \rightarrow V \int \frac{d^3\mathbf{k}}{(2\pi)^3} \, , \tag{2.115}$$

with V the volume, and taking into account that $\omega_k = ck$, one can write the energy density $\mathcal{E} = \langle \hat{H} \rangle_T / V$ as

$$\mathcal{E} = 2 \int \frac{d^3\mathbf{k}}{(2\pi)^3} \frac{c\hbar k}{e^{\beta c\hbar k} - 1} = \frac{c\hbar}{\pi^2} \int_0^\infty dk \frac{k^3}{e^{\beta c\hbar k} - 1} , \tag{2.116}$$

where the factor 2 is due to the two possible polarizations ($s = 1, 2$). By using $\omega = ck$ instead of k as integration variable one gets

$$\mathcal{E} = \frac{\hbar}{\pi^2 c^3} \int_0^\infty d\omega \frac{\omega^3}{e^{\beta\hbar\omega} - 1} = \int_0^\infty d\omega \, \rho(\omega) , \tag{2.117}$$

where

$$\rho(\omega) = \frac{\hbar}{\pi^2 c^3} \frac{\omega^3}{e^{\beta\hbar\omega} - 1} \tag{2.118}$$

is the energy density per frequency, i.e. the familiar formula of the black-body radiation, obtained for the first time in 1900 by Max Planck. The previous integral can be explicitly calculated and it gives

$$\mathcal{E} = \frac{\pi^2 k_B^4}{15 c^3 \hbar^3} T^4 , \tag{2.119}$$

which is nothing but the Stefan-Boltzmann law. In an similar way one determines the average number density of photons:

$$n = \frac{\langle \hat{N} \rangle_T}{V} = \frac{1}{\pi^2 c^3} \int_0^\infty d\omega \, \frac{\omega^2}{e^{\beta \hbar \omega} - 1} = \frac{2\zeta(3) k_B^3}{\pi^2 c^3 \hbar^3} T^3 . \tag{2.120}$$

where $\zeta(3) \simeq 1.202$. Notice that both energy density $\mathcal{E}$ and number density n of photons go to zero as the temperature T goes to zero. To conclude this section, we stress that these results are obtained at thermal equilibrium and under the condition of a vanishing chemical potential, meaning that the number of photons is not conserved when the temperature is varied.

2.4   Phase Operators

We have seen that the ladder operators $\hat{a}$ and $\hat{a}^+$ of a single mode of the electromagnetic field satisfy the fundamental relations

$$\hat{a}|n\rangle = \sqrt{n}\, |n - 1\rangle , \tag{2.121}$$
$$\hat{a}^+|n\rangle = \sqrt{n + 1}\, |n + 1\rangle , \tag{2.122}$$

where $|n\rangle$ is the Fock state, eigenstate of the number operator $\hat{N} = \hat{a}^+ \hat{a}$ and describing n photons in the mode. Remarkably, these operator can be expressed in terms of the Fock states as follows

$$\hat{a} = \sum_{n=0}^{\infty} \sqrt{n + 1}|n\rangle \langle n + 1| = |0\rangle\langle 1| + \sqrt{2}|1\rangle\langle 2| + \sqrt{3}|2\rangle\langle 3| + \cdots , \tag{2.123}$$

$$\hat{a}^+ = \sum_{n=0}^{\infty} \sqrt{n + 1}|n + 1\rangle \langle n| = |1\rangle\langle 0| + \sqrt{2}|2\rangle\langle 1| + \sqrt{3}|3\rangle\langle 2| + \cdots . \tag{2.124}$$

It is in fact straightforward to verify that the expressions (2.123) and (2.124) imply Eqs. (2.121) and (2.122).

We now introduce the phase operators

$$\hat{f} = \hat{a}\, \hat{N}^{-1/2} , \qquad \hat{f}^+ = \hat{N}^{-1/2}\, \hat{a}^+ , \tag{2.125}$$

which can be expressed as

$$\hat{f} = \sum_{n=0}^{\infty} |n\rangle\langle n+1| = |0\rangle\langle 1| + |1\rangle\langle 2| + |2\rangle\langle 3| + \cdots , \qquad (2.126)$$

$$\hat{f}^+ = \sum_{n=0}^{\infty} |n+1\rangle\langle n| = |1\rangle\langle 0| + |2\rangle\langle 1| + |3\rangle\langle 2| + \cdots , \qquad (2.127)$$

and satisfy the very nice formulas

$$\hat{f}|n\rangle = |n-1\rangle , \qquad (2.128)$$
$$\hat{f}^+|n\rangle = |n+1\rangle , \qquad (2.129)$$

showing that the phase operators act as lowering and raising operators of Fock states without the complication of coefficients in front of the obtained states.

The shifting property of the phase operators on the Fock states $|n\rangle$ resembles that of the unitary operator $e^{i\hat{p}/\hbar}$ acting on the position state $|x\rangle$ of a 1D particle, with $\hat{p}$ the 1D linear momentum which is canonically conjugated to the 1D position operator $\hat{x}$. In particular, one has

$$e^{i\hat{p}/\hbar}|x\rangle = |x-1\rangle , \qquad (2.130)$$
$$e^{-i\hat{p}/\hbar}|x\rangle = |x+1\rangle . \qquad (2.131)$$

Due to these formal analogies, in 1927 Fritz London and Paul Dirac independently suggested that the phase operator $\hat{f}$ can be written as

$$\hat{f} = e^{i\hat{\Theta}} , \qquad (2.132)$$

where $\hat{\Theta}$ is the angle operator related to the number operator $\hat{N} = \hat{a}^+\hat{a}$. But, contrary to $e^{i\hat{p}/\hbar}$, the operator $\hat{f}$ is not unitary because

$$\hat{f}\hat{f}^+ = \hat{1} , \qquad \hat{f}^+\hat{f} = \hat{1} - |0\rangle\langle 0| . \qquad (2.133)$$

This implies that the angle operator $\hat{\Theta}$ is not Hermitian and

$$\hat{f}^+ = e^{-i\hat{\Theta}^+} . \qquad (2.134)$$

In addition, a coherent state $|\alpha\rangle$ is not an eigenstate of the phase operator $\hat{f}$. However, one easily finds

$$\langle\alpha|\hat{f}|\alpha\rangle = \alpha e^{-|\alpha|^2} \sum_{n=0}^{\infty} \frac{|\alpha|^{2n}}{\sqrt{n!(n+1)!}} , \qquad (2.135)$$

and observing that for $x \gg 1$ one gets $\sum_{n=0}^{\infty} \frac{x^n}{\sqrt{n!(n+1)!}} \simeq \frac{e^x}{\sqrt{x}}$ this implies

$$\langle \alpha | \hat{f} | \alpha \rangle \simeq \frac{\alpha}{|\alpha|} = e^{i\theta} , \qquad (2.136)$$

for $\bar{N} = |\alpha|^2 \gg 1$, with $\alpha = \sqrt{\bar{N}} e^{i\theta}$. Thus, for a large average number $\bar{N}$ of photons the expectation value of the phase operator $\hat{f}$ on the coherent state $|\alpha\rangle$ gives the phase factor $e^{i\theta}$ of the complex eigenvalue α of the coherent state.

2.5 Solved Problems

Problem 2.1
Show that the eigenvalues n of the operator $\hat{N} = \hat{a}^+ \hat{a}$ are non negative.

Solution
The eigenvalue equation of $\hat{N}$ reads

$$\hat{a}^+ \hat{a} |n\rangle = n|n\rangle .$$

We can then write

$$\langle n | \hat{a}^+ \hat{a} | n \rangle = n\langle n | n \rangle = n ,$$

because the eigenstate $|n\rangle$ is normalized to one. On the other hand, we have also

$$\langle n | \hat{a}^+ \hat{a} | n \rangle = (\hat{a}|n\rangle)^+ (\hat{a}|n\rangle) = |(\hat{a}|n\rangle)|^2$$

Consequently, we get

$$n = |(\hat{a}|n\rangle)|^2 \geq 0 .$$

We stress that we have not used the commutations relations of $\hat{a}$ and $\hat{a}^+$. Thus we have actually proved that any operator $\hat{N}$ given by the factorization of a generic operator $\hat{a}$ with its self-adjunct $\hat{a}^+$ has a non negative spectrum.

Problem 2.2
Consider the operator $\hat{N} = \hat{a}^+ \hat{a}$, where $\hat{a}$ and $\hat{a}^+$ satisfy the commutation rule $\hat{a}\hat{a}^+ - \hat{a}^+ \hat{a} = 1$. Show that if $|n\rangle$ is an eigenstate of $\hat{N}$ with eigenvalue n then $a|n\rangle$ is an eigenstate of $\hat{N}$ with eigenvalue $n - 1$ and $a^+|n\rangle$ is an eigenstate of $\hat{N}$ with eigenvalue $n + 1$.

Solution
We have

$$\hat{N}\hat{a}|n\rangle = (\hat{a}^+ \hat{a})\hat{a}|n\rangle .$$

The commutation relation between $\hat{a}$ and $\hat{a}^+$ can be written as

$$\hat{a}^+\hat{a} = \hat{a}\hat{a}^+ - 1 .$$

This implies that

$$\hat{N}\hat{a}|n\rangle = (\hat{a}\hat{a}^+ - 1)\hat{a}|n\rangle = \hat{a}(\hat{a}^+\hat{a} - 1)|n\rangle = \hat{a}(\hat{N} - 1)|n\rangle = (n - 1)\hat{a}|n\rangle .$$

Finally, we obtain

$$\hat{N}\hat{a}^+|n\rangle = (\hat{a}^+\hat{a})\hat{a}^+|n\rangle = \hat{a}^+(\hat{a}\hat{a}^+)|n\rangle = \hat{a}^+(\hat{a}^+\hat{a} + 1)|n\rangle = \hat{a}^+(\hat{N} + 1)|n\rangle = (n + 1)\hat{a}^+|n\rangle .$$

Problem 2.3

Taking into account the results of the two previous problems, show that the spectrum of the number operator $\hat{N} = \hat{a}^+\hat{a}$, where $\hat{a}$ and $\hat{a}^+$ satisfy the commutation rule $\hat{a}\hat{a}^+ - \hat{a}^+\hat{a} = 1$, is the set of integer numbers.

Solution

We have seen that $\hat{N}$ has a non negative spectrum. This means that $\hat{N}$ possesses a lowest eigenvalue n_0 with $|n_0\rangle$ its eigenstate. This eigenstate $|n_0\rangle$ is such that

$$\hat{a}|n_0\rangle = 0 .$$

In fact, on the basis of the results of the previous problem, $\hat{a}|n_0\rangle$ should be eigenstate of $\hat{N}$ with eigenvalue $n_0 - 1$ but this is not possible because n_0 is the lowest eigenvalue of $\hat{N}$. Consequently the state $\hat{a}|n_0\rangle$ is not a good Fock state and we set it equal to 0. In addition, due to the fact that

$$\begin{aligned}\hat{N}|n_0\rangle &= n_0|n_0\rangle \\ &= \hat{a}^+\hat{a}|n_0\rangle = \hat{a}^+\left(\hat{a}|n_0\rangle\right) = \hat{a}^+(0) = 0\end{aligned}$$

we find that $n_0 = 0$. Thus, the state $|0\rangle$, called vacuum state, is the eigenstate of $\hat{N}$ with eigenvalue 0, i.e.

$$\hat{N}|0\rangle = 0|0\rangle = 0 ,$$

but also

$$\hat{a}|0\rangle = 0 .$$

Due to this equation, it follows that the eigenstates of $\hat{N}$ are only those generated by applying m times the operator $\hat{a}^+$ on the vacuum state $|0\rangle$, namely

$$|m\rangle = \frac{1}{\sqrt{m!}}(\hat{a}^+)^m|0\rangle ,$$

where the factorial is due to the normalization. Finally, we notice that it has been shown in the previous problem that the state $|m\rangle$ has integer eigenvalue m.

Problem 2.4

Consider the following quantum Hamiltonian of the one-dimensional harmonic oscillator

$$\hat{H} = \frac{\hat{p}^2}{2m} + \frac{1}{2}m\omega^2\hat{x}^2 \ .$$

By using the properties of the annihilation operator

$$\hat{a} = \sqrt{\frac{m\omega}{2\hbar}} \left(\hat{x} + \frac{i}{m\omega}\hat{p} \right) \ ,$$

determine the eigenfunction of the ground state of the system.

Solution

Let us observe the following property of the annihilation operator

$$\langle x|\hat{a} = \sqrt{\frac{m\omega}{2\hbar}} \left(x + \frac{\hbar}{m\omega}\frac{\partial}{\partial x} \right) \langle x| \ .$$

In addition, from

$$\hat{a}|0\rangle = 0 \ ,$$

we find

$$\langle x|\hat{a}|0\rangle = \sqrt{\frac{m\omega}{2\hbar}} \left(x + \frac{\hbar}{m\omega}\frac{\partial}{\partial x} \right) \langle x|0\rangle = 0 \ .$$

Introducing the characteristic harmonic length

$$l_H = \sqrt{\frac{\hbar}{m\omega}} \ ,$$

the dimensionless coordinate

$$\bar{x} = \frac{x}{l_H} \ ,$$

and the dimensionless eigenfunction

$$\bar{\psi}_n(\bar{x}) = l_H \ \psi_n(\bar{x}l_H) \ ,$$

we obtain

$$\left(\bar{x} + \frac{\partial}{\partial\bar{x}} \right) \bar{\psi}_0(\bar{x}) = 0 \ .$$

The solution of this first order differential equation is found by separation of variables:

$$\bar{x}\, d\bar{x} = -\frac{d\bar{\psi}_0}{\bar{\psi}_0} \,,$$

from which

$$\bar{\psi}_0(\bar{x}) = \frac{1}{\pi^{1/4}} \exp\left(\frac{-\bar{x}^2}{2}\right) ,$$

having imposed the normalization

$$\int d\bar{x}\, |\bar{\psi}_0(\bar{x})|^2 = 1 .$$

Problem 2.5
Consider the quantum Hamiltonian of the two-dimensional harmonic oscillator

$$\hat{H} = \frac{\hat{p}_1^2 + \hat{p}_2^2}{2m} + \frac{1}{2}m\omega^2(\hat{x}_1^2 + \hat{x}_2^2) .$$

By using the properties of the creation operators

$$\hat{a}_k^+ = \sqrt{\frac{m\omega}{2\hbar}}\left(\hat{x}_k - \frac{i}{m\omega}\hat{p}_k\right) , \quad k = 1, 2$$

determine the eigenfunctions of the quantum Hamiltonian with eigenvalue $6\hbar\omega$.

Solution
The eigenvalues of the two-dimensional harmonic oscillator are given by

$$E_{n_1 n_2} = \hbar\omega(n_1 + n_2 + 1) ,$$

where $n_1, n_2 = 0, 1, 2, 3, \ldots$ are the quantum numbers.
The eigenstates $|n_1 n_2\rangle$ corresponding to the eigenvalue $6\hbar\omega$ are:

$$|50\rangle , \quad |41\rangle , \quad |32\rangle , \quad |23\rangle , \quad |14\rangle , \quad |05\rangle .$$

Thus the eigenfunctions to be determined are

$$\phi_{50}(x_1, x_2) , \quad \phi_{41}(x_1, x_2) , \quad \phi_{32}(x_1, x_2) , \quad \phi_{23}(x_1, x_2) , \quad \phi_{14}(x_1, x_2) , \quad \phi_{05}(x_1, x_2) .$$

In general, the eigenfunctions $\phi_{n_1 n_2}(x_1, x_2) = \langle x_1 x_2 | n_1 n_2 \rangle$ can be factorized as follows

$$\phi_{n_1 n_2}(x_1, x_2) = \psi_{n_1}(x_1)\, \psi_{n_2}(x_2)$$

where $\psi_{n_j}(x_j) = \langle x_j | n_j \rangle$, $j = 1, 2$. It is now sufficient to calculate the following eigenfunctions of the one-dimensional harmonic oscillator:

$$\psi_0(x) , \quad \psi_1(x) , \quad \psi_2(x) , \quad \psi_3(x) , \quad \psi_4(x) , \quad \psi_5(x) .$$

We observe that the creation operator of the one-dimensional harmonic oscillator satisfies this property

$$\langle x | \hat{a}^+ = \sqrt{\frac{m\omega}{2\hbar}} \left(x - \frac{\hbar}{m\omega} \frac{\partial}{\partial x} \right) \langle x | ,$$

and consequently

$$\langle x | \hat{a}^+ | n \rangle = \sqrt{\frac{m\omega}{2\hbar}} \left(x - \frac{\hbar}{m\omega} \frac{\partial}{\partial x} \right) \langle x | n \rangle = \sqrt{\frac{m\omega}{2\hbar}} \left(x - \frac{\hbar}{m\omega} \frac{\partial}{\partial x} \right) \psi_n(x) .$$

Reminding that

$$\hat{a}^+ | n \rangle = \sqrt{n + 1} | n + 1 \rangle ,$$

we get

$$\langle x | n + 1 \rangle = \frac{1}{\sqrt{n+1}} \langle x | a^+ | n \rangle ,$$

and the iterative formula

$$\psi_{n+1}(x) = \frac{1}{\sqrt{n+1}} \sqrt{\frac{m\omega}{2\hbar}} \left(x - \frac{\hbar}{m\omega} \frac{\partial}{\partial x} \right) \psi_n(x) .$$

Now we introduce the characteristic harmonic length

$$l_H = \sqrt{\frac{\hbar}{m\omega}} ,$$

the dimensionless coordinate

$$\bar{x} = \frac{x}{l_H} ,$$

and the dimensionless wavefunction

$$\bar{\psi}_n(\bar{x}) = l_H \, \psi_n(\bar{x} l_H) ,$$

finding

$$\bar{\psi}_n(\bar{x}) = \frac{1}{\sqrt{2^n n!}} \left(\bar{x} - \frac{\partial}{\partial \bar{x}} \right)^n \bar{\psi}_0(\bar{x}) .$$

The function $\bar{\psi}_0(\bar{x})$ of the ground state is given by (see exercise 1.4)

$$\bar{\psi}_0(\bar{x}) = \frac{1}{\pi^{1/4}} \exp\left(\frac{-\bar{x}^2}{2}\right).$$

Let us now calculate the effect of the operator

$$\left(\bar{x} - \frac{\partial}{\partial \bar{x}}\right)^n.$$

For $n = 1$:

$$\left(\bar{x} - \frac{\partial}{\partial \bar{x}}\right) \exp\left(\frac{-\bar{x}^2}{2}\right) = 2\bar{x} \, \exp\left(\frac{-\bar{x}^2}{2}\right).$$

For $n = 2$:

$$\left(\bar{x} - \frac{\partial}{\partial \bar{x}}\right)^2 \exp\left(\frac{-\bar{x}^2}{2}\right) = 2(2\bar{x}^2 - 1) \, \exp\left(\frac{-\bar{x}^2}{2}\right).$$

For $n = 3$:

$$\left(\bar{x} - \frac{\partial}{\partial \bar{x}}\right)^3 \exp\left(\frac{-\bar{x}^2}{2}\right) = (8\bar{x}^3 - 12\bar{x}) \, \exp\left(\frac{-\bar{x}^2}{2}\right).$$

For $n = 4$:

$$\left(\bar{x} - \frac{\partial}{\partial \bar{x}}\right)^4 \exp\left(\frac{-\bar{x}^2}{2}\right) = (16\bar{x}^4 - 48\bar{x}^2 + 12) \, \exp\left(\frac{-\bar{x}^2}{2}\right).$$

For $n = 5$:

$$\left(\bar{x} - \frac{\partial}{\partial \bar{x}}\right)^5 \exp\left(\frac{-\bar{x}^2}{2}\right) = (32\bar{x}^5 - 160\bar{x}^3 + 120\bar{x}) \, \exp\left(\frac{-\bar{x}^2}{2}\right).$$

Finally, the eigenfunctions of H with eigenvalue $6\hbar\omega$ are linear combinations of functions $\phi_{n_1 n_2}(x_1, x_2)$, namely

$$\Phi(x_1, x_2) = \sum_{n_1 n_2} c_{n_1 n_2} \, \phi_{n_1 n_2}(x_1, x_2) \, \delta_{n_1 + n_2, 5} \,,$$

where the coefficients $c_{n_1 n_2}$ are such that

$$\sum_{n_1 n_2} |c_{n_1 n_2}|^2 \, \delta_{n_1 + n_2, 5} = 1 \,.$$

Problem 2.6

Show that the coherent state $|\alpha\rangle$, defined by the equation

$$\hat{a}|\alpha\rangle = \alpha|\alpha\rangle ,$$

can be written as

$$|\alpha\rangle = \sum_{n=0}^{\infty} e^{-|\alpha|^2/2} \frac{\alpha^n}{\sqrt{n!}}|n\rangle .$$

Solution

The coherent state $|\alpha\rangle$ can be expanded as

$$|\alpha\rangle = \sum_{n=0}^{\infty} c_n|n\rangle .$$

Then we have

$$\hat{a}|\alpha\rangle = \hat{a}\sum_{n=0}^{\infty} c_n|n\rangle = \sum_{n=0}^{\infty} c_n\hat{a}|n\rangle = \sum_{n=1}^{\infty} c_n\sqrt{n}|n-1\rangle = \sum_{n=0}^{\infty} c_{n+1}\sqrt{n+1}|n\rangle$$

$$\alpha|\alpha\rangle = \alpha\sum_{n=0}^{\infty} c_n|n\rangle = \sum_{n=0}^{\infty} \alpha\, c_n|n\rangle .$$

Since, by definition

$$\hat{a}|\alpha\rangle = \alpha|\alpha\rangle ,$$

it follows that

$$c_{n+1}\sqrt{n+1} = \alpha\, c_n$$

from which

$$c_{n+1} = \frac{\alpha}{\sqrt{n+1}}c_n ,$$

namely

$$c_1 = \frac{\alpha}{\sqrt{1}}c_0 , \quad c_2 = \frac{\alpha^2}{\sqrt{2!}}c_0 , \quad c_3 = \frac{\alpha^3}{\sqrt{3!}}c_0 , \quad \ldots$$

and in general

$$c_n = \frac{\alpha^n}{\sqrt{n!}}c_0 .$$

Summarizing

$$|\alpha\rangle = \sum_{n=0}^{\infty} c_0 \frac{\alpha^n}{\sqrt{n!}} |n\rangle \ ,$$

where the parameter c_0 is fixed by the normalization

$$1 = \langle\alpha|\alpha\rangle = |c_0|^2 \sum_{n=0}^{\infty} \frac{\alpha^{2n}}{n!} = |c_0|^2 \, e^{|\alpha|^2} \ ,$$

and consequently

$$c_0 = e^{-|\alpha|^2/2} \ ,$$

up to a constant phase factor.

Problem 2.7
Calculate the probability of finding the Fock state $|n\rangle$ in the vacuum state $|0\rangle$.

Solution
The probability p is given by

$$p = |\langle 0|n\rangle|^2 = \delta_{0,n} = \begin{cases} 1 \text{ if } n = 0 \\ 0 \text{ if } n \neq 0 \end{cases} \ .$$

Problem 2.8
Calculate the probability of finding the coherent state $|\alpha\rangle$ in the vacuum state $|0\rangle$.

Solution
The probability is given by

$$p = |\langle 0|\alpha\rangle|^2 \ .$$

Because

$$\langle 0|\alpha\rangle = \langle 0| \sum_{n=0}^{\infty} e^{-|\alpha|^2/2} \frac{\alpha^n}{\sqrt{n!}} |n\rangle = \sum_{n=0}^{\infty} e^{-|\alpha|^2/2} \frac{\alpha^n}{\sqrt{n!}} \langle 0|n\rangle = e^{-|\alpha|^2/2} \frac{1}{\sqrt{0!}} = e^{-|\alpha|^2/2} \ ,$$

we get

$$p = e^{-|\alpha|^2} \ .$$

Problem 2.9
Calculate the probability of finding the coherent state $|\alpha\rangle$ in the Fock state $|n\rangle$.

Solution

The probability is given by

$$p = |\langle n|\alpha\rangle|^2 .$$

Because

$$\langle n|\alpha\rangle = \langle n| \sum_{m=0}^{\infty} e^{-|\alpha|^2/2} \frac{\alpha^m}{\sqrt{m!}} |m\rangle = \sum_{m=0}^{\infty} e^{-|\alpha|^2/2} \frac{\alpha^m}{\sqrt{m!}} \langle n|m\rangle = e^{-|\alpha|^2/2} \frac{\alpha^n}{\sqrt{n!}} ,$$

we get

$$p = e^{-|\alpha|^2} \frac{|\alpha|^{2n}}{n!} ,$$

which is the familiar Poisson distribution.

Problem 2.10

Calculate the probability of finding the coherent state $|\alpha\rangle$ in the coherent state $|\beta\rangle$.

Solution

The probability is given by

$$p = |\langle \beta|\alpha\rangle|^2 .$$

Because

$$\langle \beta|\alpha\rangle = \langle \beta| \sum_{m=0}^{\infty} e^{-|\alpha|^2/2} \frac{\alpha^m}{\sqrt{m!}} |m\rangle = \sum_{n=0}^{\infty} \sum_{m=0}^{\infty} e^{-|\beta|^2/2} e^{-|\alpha|^2/2} \frac{(\beta^*)^n}{\sqrt{n!}} \frac{\alpha^m}{\sqrt{m!}} \langle n|m\rangle$$

$$= e^{-(|\alpha|^2+|\beta|^2)/2} \sum_{n=0}^{\infty} \frac{(\beta^*\alpha)^n}{n!} = e^{-(|\alpha|^2+|\beta|^2)/2} e^{\beta^*\alpha} = e^{-(|\alpha|^2+|\beta|^2-2\alpha\beta^*)/2} ,$$

we get

$$p = e^{-|\alpha-\beta|^2} .$$

This means that two generic coherent states $|\alpha\rangle$ and $|\beta\rangle$ are never orthogonal to each other.

Further Reading

For the second quantization of the electromagnetic field:

F. Mandl and G. Shaw, Quantum Field Theory, Chap. 1, Sects. 1.1 and 1.2 (Wiley, New York, 1984).

M.O. Scully and M.S. Zubairy, Quantum Optics, Chap. 1, Sects. 1.1 and 1.2 (Cambridge Univ. Press, Cambridge, 1997).

U. Leonhardt, Measuring the Quantum State of Light, Chap. 2, Sects. 2.1, 2.2, and 2.3 (Cambridge Univ. Press, Cambridge, 1997).

F.T. Arecchi, Measurement of the statistical distribution of Gaussian and laser sources, Phys. Rev. Lett. **15**, 912 (1965).

For the Casimir effect:

R.W. Robinett, Quantum Mechanics: Classical Results, Modern Systems, and Visualized Examples, Chap. 19, Sect. 19.9 (Oxford Univ. Press, Oxford, 2006).

S. K. Lamoreaux, Demonstration of the Casimir force in the 0.6 to 6 m range, Phys. Rev. Lett. **78**, 5 (1997).

G. Bressi, G. Carugno, R. Onofrio, and G. Ruoso, Measurement of the Casimir force between parallel metallic surfaces, Phys. Rev. Lett. **88**, 041804 (2002).

For the quantum radiation field at finite temperature:

K. Huang, Statistical Mechanics, Chap. 12, Sect. 12.1 (Wiley, New York, 1987).

Chapter 3
Electromagnetic Transitions

In this chapter we investigate the crucial role of the quantum electromagnetic field on the spontaneous and stimulated transitions between two atomic quantum states. After reviewing some properties of classical electrodynamics, we analyze the quantum electrodynamics within the dipole approximation. We calculate the rate of spontaneous emission, absorption, and stimulated emission and connect them with the transition coefficients introduced by Einstein. Finally, we discuss the lifetime of an atomic state and the line width of an electromagnetic transition.

3.1 Classical Electrodynamics

Let us consider a classical system composed of N particles with masses m_i, electric charges q_i, positions $\mathbf{r}_i$ and linear momenta $\mathbf{p}_i$, under the presence of an electromagnetic field. The Hamiltonian of free matter is given by

$$H_{free} = \sum_{i=1}^{N} \frac{\mathbf{p}_i^2}{2m_i} \, , \tag{3.1}$$

where $\mathbf{p}_i^2/(2m_i) = p_i^2/(2m_i)$ is the kinetic energy of i-th particle. The presence of the electromagnetic field is modelled by the Hamiltonian

$$H = H_{shift} + H_{rad} \, , \tag{3.2}$$

where

$$H_{shift} = \sum_{i=1}^{N} \frac{(\mathbf{p}_i - q_i \, \mathbf{A}_i)^2}{2m_i} + q_i \, \phi_i \, , \tag{3.3}$$

© Springer International Publishing AG 2017
L. Salasnich, *Quantum Physics of Light and Matter*, UNITEXT for Physics,
DOI 10.1007/978-3-319-52998-1_3

with $\mathbf{A}_i = \mathbf{A}(\mathbf{r}_i, t)$, $\phi_i = \phi(\mathbf{r}_i, t)$, where $\phi(\mathbf{r}, t)$ and $\mathbf{A}(\mathbf{r}, t)$ are the electromagnetic potentials and

$$H_{rad} = \int d^3\mathbf{r} \left(\frac{\varepsilon_0}{2} \mathbf{E}_T^2(\mathbf{r}, t) + \frac{1}{2\mu_0} \mathbf{B}^2(\mathbf{r}, t) \right) , \tag{3.4}$$

is the radiation Hamiltonian. We observe that by using the Hamilton equations

$$\dot{\mathbf{r}}_i = \frac{\partial H_{shift}}{\partial \mathbf{p}_i} \tag{3.5}$$

$$\dot{\mathbf{p}}_i = -\frac{\partial H_{shift}}{\partial \mathbf{r}_i} \tag{3.6}$$

on the shift Hamiltonian (3.3) it is straightforward to derive Newton equations with the Lorentz force acting on the i-th particle

$$m_i \ddot{\mathbf{r}}_i = q_i \left(-\nabla_i \phi_i - \frac{\partial \mathbf{A}_i}{\partial t} + \mathbf{v}_i \wedge (\nabla_i \wedge \mathbf{A}_i) \right) = q_i \left(\mathbf{E}_i + \mathbf{v}_i \wedge \mathbf{B}_i \right) \tag{3.7}$$

where $\mathbf{v}_i = \dot{\mathbf{r}}_i$ and, as always, the electric field $\mathbf{E}$ is obtained as

$$\mathbf{E} = -\nabla\phi - \frac{\partial \mathbf{A}}{\partial t} , \tag{3.8}$$

while the magnetic field $\mathbf{B}$ can be written as

$$\mathbf{B} = \nabla \wedge \mathbf{A} . \tag{3.9}$$

In Eq. (3.4) it appears the transverse electric field $\mathbf{E}_T$ which, in the Coulomb gauge $\nabla \cdot \mathbf{A} = 0$, is related to the total electric field $\mathbf{E}$ by the following decomposition into longitudinal $\mathbf{E}_L$ and transverse $\mathbf{E}_T$ fields

$$\mathbf{E} = \mathbf{E}_L + \mathbf{E}_T , \tag{3.10}$$

such that

$$\mathbf{E}_L = -\nabla\phi , \quad \mathbf{E}_T = -\frac{\partial \mathbf{A}}{\partial t} . \tag{3.11}$$

Remarkably, the longitudinal electric field $\mathbf{E}_L$ gives rise to the instantaneous electrostatic interaction between the charges. Indeed, Eq. (3.3) can be rewritten as

$$H_{shift} = \sum_{i=1}^{N} \frac{(\mathbf{p}_i - q_i \mathbf{A}_i)^2}{2m_i} + H_C , \tag{3.12}$$

where

$$H_C = \sum_{i=1}^{N} q_i \, \phi_i = \frac{1}{2} \sum_{i,j=1}^{N} \frac{q_i q_j}{4\pi\epsilon_0 |\mathbf{r}_i - \mathbf{r}_j|} \,, \tag{3.13}$$

is the Hamiltonian of the Coulomb interaction. The Hamiltonian H_{shift} can be further decomposed as follows

$$H_{shift} = H_{matt} + H_I \,, \tag{3.14}$$

where

$$H_{matt} = H_{free} + H_C \tag{3.15}$$

is the Hamiltonian of the self-interacting matter and

$$H_I = \sum_{i=1}^{N} \left(-\frac{q_i}{m} \mathbf{A}_i \cdot \mathbf{p}_i + \frac{q_i^2}{2m_i} \mathbf{A}_i^2 \right) \tag{3.16}$$

is the interaction Hamiltonian between matter and radiation. Combining the above results, we obtain the complete Hamiltonian

$$H = H_{shift} + H_{rad} = H_{matt} + H_{rad} + H_I \,. \tag{3.17}$$

Notice, however, that this Hamiltonian does not take into account the possible spin of particles.

3.2 Quantum Electrodynamics in the Dipole Approximation

The quantization of electrodynamics is obtained promoting the classical Hamiltonian of the system to a quantum Hamiltonian. For simplicity we consider the hydrogen atom with Hamiltonian

$$\hat{H}_{matt} = \frac{\hat{\mathbf{p}}^2}{2m} - \frac{e^2}{4\pi\epsilon_0 |\mathbf{r}|} \,, \tag{3.18}$$

where $\hat{\mathbf{p}} = -i\hbar \nabla$ is the linear momentum operator of the electron in the state $|\mathbf{p}\rangle$, in the presence of the radiation field with Hamiltonian

$$\hat{H}_{rad} = \sum_{\mathbf{k}} \sum_{s} \hbar\omega_k \, \hat{a}_{\mathbf{k}s}^{+} \hat{a}_{\mathbf{k}s} \,, \tag{3.19}$$

where $\hat{a}_{\mathbf{k}s}$ and $\hat{a}_{\mathbf{k}s}^{+}$ are the annihilation and creation operators of the photon in the state $|\mathbf{k}s\rangle$. Here $-e$ is the electric charge of the electron, $e = 1.60 \times 10^{-19}$ C is the electric charge of the proton, and $m = m_e m_p / (m_e + m_p) \simeq m_e$ is the reduced mass of the electron-proton system, with $m_e = 9.11 \times 10^{-31}$ kg the electron mass and $m_p = 1.67 \times 10^{-27}$ kg the proton mass.

We take into account the matter-radiation interaction by using the so-called dipolar Hamiltonian

$$\hat{H}_D = \frac{e}{m}\hat{\mathbf{A}}(\mathbf{0}, t) \cdot \hat{\mathbf{p}} . \tag{3.20}$$

With respect to the complete interaction Hamiltonian $\hat{H}_I$, given in our case by Eq. (3.16) with $N = 1$, we neglect the term $e^2\hat{\mathbf{A}}^2/(2m)$ of Eq. (3.16) which represents only a tiny perturbation in the atomic system. In addition, we also neglect the spatial variations in the vector potential operator

$$\hat{\mathbf{A}}(\mathbf{r}, t) = \sum_{\mathbf{k}}\sum_{s}\sqrt{\frac{\hbar}{2\epsilon_0\omega_k V}}\left[\hat{a}_{\mathbf{k}s}\, e^{i(\mathbf{k}\cdot\mathbf{r}-\omega_k t)} + \hat{a}_{\mathbf{k}s}^+\, e^{-i(\mathbf{k}\cdot\mathbf{r}-\omega_k t)}\right] \boldsymbol{\varepsilon}_{\mathbf{k}s} . \tag{3.21}$$

This dipolar approximation, which corresponds to

$$e^{i\mathbf{k}\cdot\mathbf{r}} = 1 + i\mathbf{k}\cdot\mathbf{r} + \frac{1}{2}(i\mathbf{k}\cdot\mathbf{r})^2 + \cdots \simeq 1 , \tag{3.22}$$

$$e^{-i\mathbf{k}\cdot\mathbf{r}} = 1 - i\mathbf{k}\cdot\mathbf{r} + \frac{1}{2}(i\mathbf{k}\cdot\mathbf{r})^2 + \cdots \simeq 1 , \tag{3.23}$$

is reliable if $\mathbf{k}\cdot\mathbf{r} \ll 1$, namely if the electromagnetic radiation has a wavelength $\lambda = 2\pi/|\mathbf{k}|$ very large compared to the linear dimension R of the atom. Indeed, the approximation is fully justified in atomic physics where $\lambda \simeq 10^{-7}$ m and $R \simeq 10^{-10}$ m. Notice, however, that for the γ electromagnetic transitions of atomic nuclei the dipolar approximation is not good and it is more convenient to expand the vector potential operator $\hat{\mathbf{A}}(\mathbf{r}, t)$ into vector spherical harmonics, i.e. photons of definite angular momentum.

The total Hamiltonian of our system is then given by

$$\hat{H} = \hat{H}_0 + \hat{H}_D , \tag{3.24}$$

where

$$\hat{H}_0 = \hat{H}_{matt} + \hat{H}_{rad} \tag{3.25}$$

is the unperturbed Hamiltonian, whose eigenstates are of the form

$$|a\rangle|\ldots n_{\mathbf{k}s}\ldots\rangle = |a\rangle \otimes |\ldots n_{\mathbf{k}s}\ldots\rangle \tag{3.26}$$

where $|a\rangle$ is the eigenstate of $\hat{H}_{matt}$ with eigenvalue E_a and $|\ldots n_{\mathbf{k}s}\ldots\rangle$ it the eigenstate of $\hat{H}_{rad}$ with eigenvalue $\sum_{\mathbf{k}s}\hbar\omega_k n_{\mathbf{k}s}$, i.e.

$$\hat{H}_0 |a\rangle |\ldots n_{\mathbf{k}r} \ldots\rangle = \left(\hat{H}_{matt} + \hat{H}_{rad}\right) |a\rangle |\ldots n_{\mathbf{k}s} \ldots\rangle = \left(E_a + \sum_{\mathbf{k}s} \hbar\omega_k n_{\mathbf{k}s}\right) |a\rangle |\ldots n_{\mathbf{k}s} \ldots\rangle .$$

(3.27)

The time-dependent perturbing Hamiltonian is instead given by

$$\hat{H}_D(t) = \frac{e}{m} \sum_{\mathbf{k}} \sum_s \sqrt{\frac{\hbar}{2\epsilon_0 \omega_k V}} \left[\hat{a}_{\mathbf{k}s} e^{-i\omega_k t} + \hat{a}_{\mathbf{k}s}^+ e^{i\omega_k t}\right] \varepsilon_{\mathbf{k}s} \cdot \hat{\mathbf{p}} .$$

(3.28)

Fermi golden rule: Given the initial $|I\rangle$ and final $|F\rangle$ eigenstates of the unperturbed Hamiltonian $\hat{H}_0$ under the presence to the time-dependent perturbing Hamiltonian $\hat{H}_D$, the probability per unit time of the transition from $|I\rangle$ to $|F\rangle$ is given by

$$W_{IF} = \frac{2\pi}{\hbar} |\langle F|\hat{H}_D(0)|I\rangle|^2 \, \delta(E_I - E_F) ,$$

(3.29)

with the constraint of energy conservation.

This is the so-called Fermi golden rule, derived in 1926 by Paul Dirac on the basis of the first order time-dependent perturbation theory, and named "golden rule" few years later by Enrico Fermi.

3.2.1 Spontaneous Emission

Let us now apply the Fermi golden rule to the very interesting case of the hydrogen atom in the state $|b\rangle$ and the radiation field in the vacuum state $|0\rangle$. We are thus supposing that the initial state is

$$|I\rangle = |b\rangle |0\rangle .$$

(3.30)

Notice that, because we are considering the hydrogen atom, one has

$$\hat{H}_{matt}|b\rangle = E_b|b\rangle ,$$

(3.31)

where

$$E_b = -\frac{mc^2\alpha^2}{2n_b^2} \simeq -\frac{13.6 \text{ eV}}{n_b^2}$$

(3.32)

is the well-known quantization formula of the nonrelativistic hydrogen atom with quantum number $n_b = 1, 2, 3, \ldots$ and

$$\alpha = \frac{e^2}{(4\pi\epsilon_0)\hbar c} \simeq \frac{1}{137} \tag{3.33}$$

the fine-structure constant. In addition we suppose that the final state is

$$|F\rangle = |a\rangle|\mathbf{k}s\rangle , \tag{3.34}$$

i.e. the final atomic state is $|a\rangle$ and the final photon state is $|\mathbf{k}s\rangle = |1_{\mathbf{k}s}\rangle = \hat{a}_{\mathbf{k}s}^+|0\rangle$. Obviously, we have

$$\hat{H}_{matt}|a\rangle = E_a|a\rangle , \tag{3.35}$$

where

$$E_a = -\frac{mc^2\alpha^2}{2n_a^2} \simeq -\frac{13.6\,\text{eV}}{n_a^2} \tag{3.36}$$

with $n_a = 1, 2, 3, \ldots$. In this process, where $E_b > E_a$, there is the spontaneous production of a photon from the vacuum: a phenomenon strictly related to the quantization of the electromagnetic field.

From Eqs. (3.28) and (3.29) one finds

$$W_{ba,\mathbf{k}s}^{spont} = \frac{2\pi}{\hbar} \left(\frac{e}{m}\right)^2 \left(\frac{\hbar}{2\epsilon_0\omega_k V}\right) |\boldsymbol{\varepsilon}_{\mathbf{k}s} \cdot \langle a|\hat{\mathbf{p}}|b\rangle|^2 \, \delta(E_b - E_a - \hbar\omega_k) , \tag{3.37}$$

because

$$\hat{a}_{\mathbf{k}'s'}|I\rangle = \hat{a}_{\mathbf{k}'s'}|b\rangle|0\rangle = |b\rangle\hat{a}_{\mathbf{k}'s'}|0\rangle = 0 , \tag{3.38}$$

while

$$\hat{a}_{\mathbf{k}'s'}^+|I\rangle = \hat{a}_{\mathbf{k}'s'}^+|b\rangle|0\rangle = |b\rangle\hat{a}_{\mathbf{k}'s'}^+|0\rangle = |b\rangle|\mathbf{k}'s'\rangle , \tag{3.39}$$

and consequently

$$\langle F|\hat{\mathbf{p}}\,\hat{a}_{\mathbf{k}'s'}|I\rangle = 0 , \quad \langle F|\hat{\mathbf{p}}\,\hat{a}_{\mathbf{k}'s'}^+|I\rangle = \langle a|\hat{\mathbf{p}}|b\rangle \,\delta_{\mathbf{k}',\mathbf{k}}\,\delta_{s',s} . \tag{3.40}$$

Remember that $\hat{\mathbf{p}}$ acts on a different Hilbert space with respect to $\hat{a}_{\mathbf{k}s}$ and $\hat{a}_{\mathbf{k}s}^+$. On the basis of Heisenberg equation of motion of the linear momentum operator $\hat{\mathbf{p}}$ of the electron

$$\frac{\hat{\mathbf{p}}}{m} = \frac{d\mathbf{r}}{dt} = \frac{1}{i\hbar}[\mathbf{r}, \hat{H}_{matt}] , \tag{3.41}$$

we get

$$\begin{aligned}
\langle a|\hat{\mathbf{p}}|b\rangle &= \langle a|m\frac{1}{i\hbar}[\mathbf{r}, \hat{H}_{matt}]|b\rangle = \frac{m}{i\hbar}\langle a|\mathbf{r}\hat{H}_{matt} - \hat{H}_{matt}\mathbf{r}|b\rangle \\
&= \frac{m}{i\hbar}(E_b - E_a)\langle b|\mathbf{r}|a\rangle = -im\omega_{ba}\,\langle a|\mathbf{r}|b\rangle ,
\end{aligned} \tag{3.42}$$

where $\omega_{ba} = (E_b - E_a)/\hbar$, and consequently

$$W_{ba,ks}^{spont} = \frac{\pi\omega_{ba}^2}{V\epsilon_0\omega_k} |\varepsilon_{ks} \cdot \langle a|e\,\mathbf{r}|b\rangle|^2 \,\delta(\hbar\omega_{ba} - \hbar\omega_k) \,. \tag{3.43}$$

The delta function is eliminated by integrating over the final photon states

$$W_{ba}^{spont} = \sum_{\mathbf{k}}\sum_s W_{ba,ks}^{spont} = V\int \frac{d^3\mathbf{k}}{(2\pi)^3} \sum_{s=1,2} W_{ba,ks}^{spont} = \frac{V}{8\pi^3}\int dk\,k^2 \int d\Omega \sum_{s=1,2} W_{ba,ks}^{spont} \,, \tag{3.44}$$

where $d\Omega$ is the differential solid angle. Because $\varepsilon_{\mathbf{k}1}$, $\varepsilon_{\mathbf{k}2}$ and $\mathbf{n} = \mathbf{k}/k$ form a orthonormal system of vectors, setting $\mathbf{r}_{ab} = \langle a|\mathbf{r}|b\rangle$ one finds

$$|\mathbf{r}_{ab}|^2 = |\varepsilon_{\mathbf{k}1}\cdot\mathbf{r}_{ab}|^2 + |\varepsilon_{\mathbf{k}2}\cdot\mathbf{r}_{ab}|^2 + |\mathbf{n}\cdot\mathbf{r}_{ab}|^2 = \sum_{s=1,2} |\varepsilon_{ks}\cdot\mathbf{r}_{ab}|^2 + |\mathbf{r}_{ab}|^2\cos^2(\theta) \,, \tag{3.45}$$

where θ is the angle between $\mathbf{r}_{ba}$ and $\mathbf{n}$. It follows immediately

$$\sum_{s=1,2} |\varepsilon_{ks}\cdot\mathbf{r}_{ab}|^2 = |\mathbf{r}_{ab}|^2(1 - \cos^2(\theta)) = |\mathbf{r}_{ab}|^2\sin^2(\theta) \,, \tag{3.46}$$

namely

$$\sum_{s=1,2} |\varepsilon_{ks}\cdot\langle a|\mathbf{r}|b\rangle|^2 = |\langle a|\mathbf{r}|b\rangle|^2\sin^2(\theta) \,. \tag{3.47}$$

In addition, in spherical coordinates one can choose $d\Omega = \sin(\theta)d\theta d\phi$, with $\theta \in [0, \pi]$ the zenith angle of colatitude and $\phi \in [0, 2\pi]$ the azimuth angle of longitude, and then

$$\int d\Omega \sin^2(\theta) = \int_0^{2\pi} d\phi \int_0^\pi d\theta \sin^3(\theta) = \frac{8\pi}{3} \,. \tag{3.48}$$

In this way from Eq. (3.44) we finally obtain

$$W_{ba}^{spont} = \frac{\omega_{ba}^3}{3\pi\epsilon_0\hbar c^3} |\langle a|\mathbf{d}|b\rangle|^2 \,, \tag{3.49}$$

where the $\mathbf{d} = -e\,\mathbf{r}$ is the classical electric dipole momentum of the hydrogen atom, i.e. the dipole of the electron-proton system where $\mathbf{r}$ is the position of the electron of charge $-e < 0$ with respect to the proton of charge $e > 0$, and $\langle a|\mathbf{d}|b\rangle = -\langle a|e\,\mathbf{r}|b\rangle$ is the so-called dipole transition element.

3.2.2   Absorption

We now consider the excitation from the atomic state $|a\rangle$ to the atomic state $|b\rangle$ due to the absorption of one photon. Thus we suppose that the initial state is

$$|I\rangle = |a\rangle|n_{\mathbf{k}s}\rangle , \tag{3.50}$$

while the final state is

$$|F\rangle = |b\rangle|n_{\mathbf{k}s} - 1\rangle , \tag{3.51}$$

where $E_a < E_b$. From Eqs. (3.28) and (3.29) one finds

$$W_{ab,\mathbf{k}s}^{absorp} = \frac{2\pi}{\hbar} \left(\frac{e}{m}\right)^2 \left(\frac{\hbar}{2\epsilon_0\omega_k V}\right) n_{\mathbf{k}s} \, |\varepsilon_{\mathbf{k}s} \cdot \langle b|\hat{\mathbf{p}}|a\rangle|^2 \, \delta(E_a + \hbar\omega_k - E_b) , \tag{3.52}$$

because

$$\langle F|\hat{\mathbf{p}}\,\hat{a}_{\mathbf{k}'s'}|I\rangle = \sqrt{n_{\mathbf{k}s}} \, \langle b|\hat{\mathbf{p}}|a\rangle \, \delta_{\mathbf{k}',\mathbf{k}} \, \delta_{s',s} , \quad \langle F|\hat{\mathbf{p}}\,\hat{a}_{\mathbf{k}'s'}^+|I\rangle = 0 . \tag{3.53}$$

Note that with respect to Eq. (3.37) in Eq. (3.52) there is also the multiplicative term $n_{\mathbf{k}s}$. We can follow the procedure of the previous section to get

$$W_{ab,\mathbf{k}s}^{absorp} = \frac{\pi\omega_{ba}^2}{V\,\epsilon_0\omega_k} \, n_{\mathbf{k}s} \, |\varepsilon_{\mathbf{k}s} \cdot \langle b|e\,\mathbf{r}|a\rangle|^2 \, \delta(\hbar\omega_{ba} - \hbar\omega_k) . \tag{3.54}$$

Again the delta function can be eliminated by integrating over the final photon states but here one must choose the functional dependence of $n_{\mathbf{k}s}$. We simply set

$$n_{\mathbf{k}s} = n(\omega_k) , \tag{3.55}$$

and after integration over $\mathbf{k}$ and s, from Eq. (3.54) we get

$$W_{ab}^{absorp} = \frac{\omega_{ba}^3}{3\pi\epsilon_0\hbar c^3} \, |\langle b|\mathbf{d}|a\rangle|^2 \, n(\omega_{ba}) = W_{ba}^{spont} \, n(\omega_{ba}) . \tag{3.56}$$

Note that without the integration over the solid angle, which gives a factor $1/3$, one obtains the absorption probability of one photon with a specific direction. For a thermal distribution of photons, with $\rho(\omega)$ the energy density per unit of angular frequency specified by the thermal-equilibrium Planck formula

$$\rho(\omega) = \frac{\hbar\omega^3}{\pi^2 c^3} \, n(\omega) , \quad n(\omega) = \frac{1}{e^{\hbar\omega/(k_B T)} - 1} , \tag{3.57}$$

where k_B is the Boltzmann constant and T the absolute temperature, we can also write

$$W_{ab}^{absorp} = |\langle b|\mathbf{d}|a\rangle|^2 \frac{1}{3\epsilon_0\hbar^2} \rho(\omega_{ba}) = W_{ba}^{spont} \frac{\pi^2 c^3}{\hbar\omega_{ba}^3} \rho(\omega_{ba}) . \qquad (3.58)$$

3.2.3 Stimulated Emission

In conclusion we consider the stimulated emission of a photon from the atomic state $|b\rangle$ to the atomic state $|a\rangle$. Thus we suppose that the initial state is

$$|I\rangle = |b\rangle|n_{\mathbf{ks}}\rangle , \qquad (3.59)$$

while the final state is

$$|F\rangle = |a\rangle|n_{\mathbf{ks}} + 1\rangle , \qquad (3.60)$$

where $E_b > E_a$. From Eqs. (3.28) and (3.29) one finds

$$W_{ba,\mathbf{ks}}^{stimul} = \frac{2\pi}{\hbar} \left(\frac{e}{m}\right)^2 \left(\frac{\hbar}{2\epsilon_0\omega_k V}\right) (n_{\mathbf{ks}} + 1) |\boldsymbol{\varepsilon}_{\mathbf{ks}} \cdot \langle a|\hat{\mathbf{p}}|b\rangle|^2 \delta(E_b - E_a - \hbar\omega_k) , \qquad (3.61)$$

because

$$\hat{a}_{\mathbf{ks}}|I\rangle = \hat{a}_{\mathbf{ks}}|b\rangle|n_{\mathbf{ks}}\rangle = |b\rangle\hat{a}_{\mathbf{ks}}|n_{\mathbf{ks}}\rangle = \sqrt{n_{\mathbf{ks}}} |b\rangle|n_{\mathbf{ks}} - 1\rangle , \qquad (3.62)$$

while

$$\hat{a}_{\mathbf{ks}}^+|I\rangle = \hat{a}_{\mathbf{ks}}^+|b\rangle|n_{\mathbf{ks}}\rangle = |b\rangle\hat{a}_{\mathbf{ks}}^+|n_{\mathbf{ks}}\rangle = \sqrt{n_{\mathbf{ks}} + 1} |b\rangle|n_{\mathbf{ks}} + 1\rangle , \qquad (3.63)$$

and consequently

$$\langle F|\hat{\mathbf{p}}\,\hat{a}_{\mathbf{ks}}|I\rangle = 0 , \quad \langle F|\hat{\mathbf{p}}\,\hat{a}_{\mathbf{ks}}^+|I\rangle = \sqrt{n_{\mathbf{ks}} + 1} \langle a|\hat{\mathbf{p}}|b\rangle . \qquad (3.64)$$

Note that with respect to Eq. (3.52) in Eq. (3.61) there is the factor $n_{\mathbf{ks}} + 1$ instead of $n_{\mathbf{ks}}$. It is straightforward to follow the procedure of the previous section to obtain

$$W_{ba}^{stimul} = \frac{\omega_{ba}^3}{3\pi\epsilon_0\hbar c^3} |\langle a|\mathbf{d}|b\rangle|^2 (n(\omega_{ba}) + 1) = W_{ab}^{absorp} + W_{ba}^{spont} , \qquad (3.65)$$

which shows that the probability W_{ba}^{stimul} of stimulated emission reduces to the spontaneous one W_{ba}^{spont} when $n(\omega_{ba}) = 0$. It is then useful to introduce

$$\tilde{W}_{ba}^{stimul} = W_{ba}^{stimul} - W_{ba}^{spont} \qquad (3.66)$$

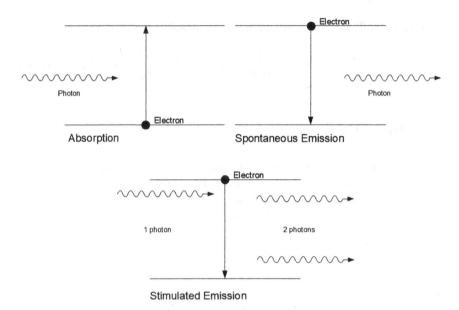

Fig. 3.1 Scheme of the 3 mechanisms of radiative transition

which is the effective stimulated emission, i.e. the stimulated emission without the contribution due to the spontaneous emission. Clearly, for a very large number of photons ($n(\omega_{ba}) \gg 1$) one gets $\tilde{W}_{ba}^{stimul} \simeq W_{ba}^{stimul}$. Moreover, for a thermal distribution of photons, with the energy density per unit of angular frequency $\rho(\omega)$, we can also write

$$\tilde{W}_{ba}^{stimul} = W_{ab}^{absorp} = W_{ba}^{spont} \frac{\pi^2 c^3}{\hbar \omega_{ba}^3} \rho(\omega_{ba}) . \tag{3.67}$$

Remarkably, the probability of stimulated emission is different from zero only if the emitted photon is in the same single-particle state $|\mathbf{k}s\rangle$ of the stimulating ones (apart when the stimulating light is the vacuum $|0\rangle$). In the stimulated emission the emitted photon is said to be "coherent" with the stimulating ones, having the same frequency and the same direction (Fig. 3.1).

3.3 Selection Rules

It is clear that, within the dipolar approximation, no electromagnetic transition, either spontaneous or stimulated, will occur between the atomic states $|a\rangle$ and $|b\rangle$ unless at least one component of the dipole transition matrix element $\langle b|\mathbf{d}|a\rangle$ is nonzero. It is possible to show that the matrix elements are zero for certain pairs of states. If so, the transition is not allowed (at least in the electric dipole approximation), and the results can be summarized in terms of simple selection rules governing the allowed changes in quantum numbers in transitions.

Since the electric dipole $\mathbf{d} = -e\mathbf{r}$ changes sign under parity ($\mathbf{r} \rightarrow -\mathbf{r}$), the matrix element $\langle b|\mathbf{d}|a\rangle$ is zero if the states $|a\rangle$ and $|b\rangle$ have the same parity. Therefore

the parity of the wavefunction must change in an electric dipole transition.

This means that if $\psi_a(\mathbf{r})$ and $\psi_b(\mathbf{r})$ are the eigenfunctions of the states $|a\rangle$ and $|b\rangle$ and, for instance, both eigenfunctions are even, i.e.

$$\psi_a(-\mathbf{r}) = \psi_a(\mathbf{r}) \quad \text{and} \quad \psi_b(-\mathbf{r}) = \psi_b(\mathbf{r}) , \tag{3.68}$$

then the dipole matrix element is such that

$$\langle b|\mathbf{d}|a\rangle = -e \int d^3\mathbf{r} \; \psi_b^*(\mathbf{r}) \; \mathbf{r} \; \psi_a(\mathbf{r}) = \mathbf{0} . \tag{3.69}$$

In fact, under the transformation $\mathbf{r} \rightarrow -\mathbf{r}$ one finds $\int d^3(-\mathbf{r}) = \int d^3\mathbf{r}$ while $\langle b|\mathbf{d}|a\rangle = -\langle b|\mathbf{d}|a\rangle$, and then it follows $\langle b|\mathbf{d}|a\rangle = \mathbf{0}$.

Let us recall that in our calculations a generic eigenstate $|a\rangle = |nlm\rangle$ of the matter Hamiltonian $\hat{H}_{matt}$ is such that

$$\psi_a(\mathbf{r}) = \psi_{nlm}(\mathbf{r}) = \langle \mathbf{r}|a\rangle = \langle \mathbf{r}|nlm\rangle = R_{nl}(r)Y_{lm}(\theta, \phi) , \tag{3.70}$$

with $Y_{nm}(\theta, \phi)$ the spherical harmonic function. The spherical harmonics satisfy the orthonormalization condition

$$\int_0^\pi d\theta \; \sin(\theta) \int_0^{2\pi} d\phi \; Y_{l'm'}^*(\theta, \phi) \; Y_{lm}(\theta, \phi) = \delta_{l,l'} \; \delta_{m,m'} , \tag{3.71}$$

where $Y_{l,m}^*(\theta, \phi) = (-1)^m Y_{l,-m}(\theta, \phi)$. Moreover under parity transformation one finds

$$\psi_{nlm}(-\mathbf{r}) = R_{nl}(r)Y_{lm}(\pi - \theta, \phi + \pi) = (-1)^l R_{nl}(r)Y_{lm}(\theta, \phi) = (-1)^l \psi_{nlm}(\mathbf{r}) , \tag{3.72}$$

thus the radial part $R_{nl}(r)$ of the wavefunction is unchanged and the parity of the state is fully determined from the angular part.

The generic dipole matrix element is given by

$$\langle n'l'm'|\mathbf{d}|nlm\rangle = -e \int_0^\infty dr \; r^3 R_{n'l'}(r)R_{nl}(r) \int_0^\pi d\theta \; \sin(\theta) \int_0^{2\pi} d\phi \; Y_{l'm'}^*(\theta, \phi)$$
$$\times \; (\cos(\phi) \; \sin(\theta), \sin(\phi) \; \sin(\theta), \cos(\theta)) \; Y_{lm}(\theta, \phi) , \tag{3.73}$$

with $\mathbf{r} = r\,(\cos(\phi)\,\sin(\theta), \sin(\phi)\,\sin(\theta), \cos(\theta))$. Setting $\Delta l = l - l'$ and $\Delta m = m - m'$, by using the properties of the spherical harmonics it is possible to prove that the dipole matrix element $\langle n'l'm'|\mathbf{d}|nlm\rangle$ is different from zero if only if $\Delta l = \pm 1$ and $\Delta m = 0, \pm 1$ (see, for instance, Problem 3.2). Thus, in the dipole approximation, the orbital angular momentum l of the atomic state and its third component $m = -l, -l+1, \ldots, l-1, l$ must satisfy the selection rules

$$\Delta l = \pm 1 \quad \Delta m = 0, \pm 1 . \tag{3.74}$$

This means that in the electric dipole transitions the photon carries off (or brings in) one unit of angular momentum. It is important to stress that the above selection rules are obtained within the dipole approximation. This means that they can be violated by rare electromagnetic transitions involving higher multipolarities.

Concluding this section we observe that in the general case of N particles with positions $\mathbf{r}_i$ and electric charges q_i, $i = 1, 2, \ldots, N$, the electric dipole momentum is defined

$$\mathbf{d} = \sum_{i=1}^{N} q_i\, \mathbf{r}_i . \tag{3.75}$$

Moreover, due to Eq. (3.42), the dipolar interaction Hamiltonian (3.20) can be effectively written as

$$\hat{H}_D = -\mathbf{d} \cdot \hat{\mathbf{E}} , \tag{3.76}$$

which is the familiar expression of the coupling between the electric dipole $\mathbf{d}$ and the electric field $\hat{\mathbf{E}}$.

3.4   Einstein Coefficients

In 1919 Albert Einstein observed that, given an ensemble of N atoms in two possible atomic states $|a\rangle$ and $|b\rangle$, with $N_a(t)$ the number of atoms in the state $|a\rangle$ at time t and $N_b(t)$ the number of atoms in the state $|b\rangle$ at time t, it must be

$$N = N_a(t) + N_b(t) \tag{3.77}$$

and consequently

$$\frac{dN_a}{dt} = -\frac{dN_b}{dt} . \tag{3.78}$$

Einstein suggested that, if the atoms are exposed to an electromagnetic radiation of energy density per unit of frequency $\rho(\omega)$, the rate of change of atoms in the state $|a\rangle$ must be

$$\frac{dN_a}{dt} = A_{ba}\, N_b + B_{ba}\, \rho(\omega_{ba})\, N_b - B_{ab}\, \rho(\omega_{ba})\, N_a . \tag{3.79}$$

where the parameters A_{ba}, B_{ba}, and B_{ab} are known as Einstein coefficients. The first two terms in this equation describe the increase of the number of atoms in $|a\rangle$ due to spontaneous and stimulated transitions from $|b\rangle$, while the third term takes into account the decrease of the number of atoms in $|a\rangle$ due to absorption with consequent transition to $|b\rangle$.

Einstein was able to determine the coefficients A_{ba}, B_{ba} and B_{ab} by supposing that the two rates in Eqs. (3.78) and (3.79) must be equal to zero at thermal equilibrium, i.e.

$$\frac{dN_a}{dt} = -\frac{dN_b}{dt} = 0 , \tag{3.80}$$

In this way Einstein found

$$A_{ba}\frac{N_b}{N_a} = \rho(\omega_{ba})\left(B_{ab} - B_{ba}\frac{N_b}{N_a}\right) . \tag{3.81}$$

Because the relative population of the atomic states $|a\rangle$ and $|b\rangle$ is given by a Boltzmann factor

$$\frac{N_b}{N_a} = \frac{e^{-\beta E_b}}{e^{-\beta E_a}} = e^{-\beta(E_b - E_a)} = e^{-\beta\hbar\omega_{ba}} , \tag{3.82}$$

Einstein got

$$\rho(\omega_{ba}) = \frac{A_{ba}}{B_{ab}e^{\beta\hbar\omega_{ba}} - B_{ba}} . \tag{3.83}$$

At thermal equilibrium we know that

$$\rho(\omega_{ba}) = \frac{\hbar\omega_{ba}^3}{\pi^2 c^3}\frac{1}{e^{\beta\hbar\omega_{ba}} - 1} . \tag{3.84}$$

It follows that

$$A_{ba} = B_{ba}\frac{\hbar\omega_{ba}^3}{\pi^2 c^3} , \quad B_{ab} = B_{ba} . \tag{3.85}$$

Notice that in this way Einstein obtained the coefficient A_{ba} of spontaneous decay by simply calculating the coefficient of stimulated decay B_{ba}.

Actually, by using the results we have obtained in the previous section, it is clear that

$$A_{ba} = W_{ba}^{spont} = \frac{\omega_{ba}^3}{3\pi\epsilon_0\hbar c^3}|\langle a|\mathbf{d}|b\rangle|^2 , \tag{3.86}$$

$$B_{ba} = \frac{\tilde{W}_{ba}^{stimul}}{\rho(\omega_{ba})} = \frac{\pi^2 c^3}{\hbar\omega_{ba}^3}A_{ba} , \tag{3.87}$$

$$B_{ab} = \frac{W_{ab}^{absorp}}{\rho(\omega_{ba})} = \frac{\pi^2 c^3}{\hbar\omega_{ba}^3}A_{ba} . \tag{3.88}$$

It is important to stress that the laser device, invented in 1957 by Charles Townes and Arthur Schawlow at Bell Labs, is based on a generalization of Eqs. (3.78) and (3.79). In the laser device one has population inversion, which is achieved with an out-of-equilibrium pumping mechanism strongly dependent on the specific characteristics of the laser device.

3.4.1 Rate Equations for Two-Level and Three-Level Systems

Let us analyze under which conditions one can get population inversion. We have seen that for a two-level system under the action of an electromagnetic pump with energy density $\rho(\omega_{ba})$ the rate equations of Einstein are given by

$$\frac{dN_a}{dt} = A_{ba} N_b + R (N_b - N_a) , \tag{3.89}$$

$$\frac{dN_b}{dt} = -A_{ba} N_b - R (N_b - N_a) , \tag{3.90}$$

where $R = B_{ba}\rho(\omega_{ba})$ is the pumping rate. We now suppose that the system is in a steady state, but not at thermal equilibrium, i.e. the energy density $\rho(\omega_{ba})$ of the electromagnetic pump is not given by the Planck distribution (3.84). At the steady state we have

$$\frac{dN_a}{dt} = \frac{dN_b}{dt} = 0 , \tag{3.91}$$

from which, taking into account that $N = N_a + N_b$, we obtain

$$N_a = N \frac{A_{ba} + R}{A_{ba} + 2R} , \tag{3.92}$$

$$N_b = N \frac{R}{A_{ba} + 2R} , \tag{3.93}$$

and the population imbalance $\zeta = (N_a - N_b)/N$ reads

$$\zeta = \frac{A_{ba}}{A_{ba} + 2R} . \tag{3.94}$$

Equation (3.94) shows that in the absence of electromagnetic pump ($R = 0$) one has $\zeta = 1$, which simply means that all two-level atoms are in the lower state $|a\rangle$. Instead, if the electromagnetic pump is active ($R \neq 0$) the population imbalance ζ decreases by increasing the pumping rate R, but ζ cannot change sign because $\zeta \to 0$ for $R \to +\infty$. Thus, we conclude that the population inversion ($\zeta < 0$), and consequently laser light, is impossible in a strictly two-level system.

Let us now consider an ensemble of N atoms in three possible atomic states $|a\rangle$, $|b\rangle$ and $|c\rangle$, with $N_a(t)$ the number of atoms in the state $|a\rangle$ at time t, $N_b(t)$ the number of atoms in the state $|b\rangle$ at time t, and $N_c(t)$ the number of atoms in the state $|b\rangle$ at time t. In this ensemble of three-level atoms it must be

$$N = N_a(t) + N_b(t) + N_c(t) \ . \tag{3.95}$$

Moreover, we suppose that $E_a < E_b < E_c$, with E_a the energy of the state $|a\rangle$, E_b the energy of the state $|b\rangle$, and E_c the energy of the state $|c\rangle$. In the presence of an electromagnetic pump of energy density $\rho(\omega_{ca})$, with $\omega_{ca} = (E_c - E_a)/\hbar$, between the state $|a\rangle$ and $|c\rangle$ the rate equations of the three-level system are given by

$$\frac{dN_a}{dt} = A_{ba}\,N_b + A_{ca}\,N_c + R\,(N_c - N_a) + B_{ba}\,\rho_{ba}\,(N_b - N_a) \ , \tag{3.96}$$

$$\frac{dN_b}{dt} = -A_{ba}\,N_b + A_{cb}N_c - B_{ba}\,\rho_{ba}\,(N_b - N_a) + B_{cb}\,\rho_{cb}\,(N_c - N_b) \ , \tag{3.97}$$

$$\frac{dN_c}{dt} = -A_{cb}\,N_c - A_{ca}\,N_c - R\,(N_c - N_a) - B_{cb}\,\rho_{cb}\,(N_c - N_b) \ , \tag{3.98}$$

where $R = B_{ca}\,\rho(\omega_{ca})$ is the pumping rate and the energy densities ρ_{ba} and ρ_{cb} are induced by the electromagnetic transitions. These equations are simplified under the following physical assumptions (Ruby laser): (i) there is no electromagnetic transition from $|c\rangle$ and $|b\rangle$, namely $\rho_{cb} = 0$; (ii) the time decay of the spontaneous transition from $|c\rangle$ to $|a\rangle$ is very long, namely $A_{ca} \simeq 0$; (iii) the time decay of the spontaneous transition from $|c\rangle$ to $|b\rangle$ is very short and consequently $N_c \ll N_a$. In this way the rate equations become

$$\frac{dN_a}{dt} = A_{ba}\,N_b - R\,N_a + S\,(N_b - N_a) \ , \tag{3.99}$$

$$\frac{dN_b}{dt} = -A_{ba}\,N_b + A_{cb}N_c - S\,(N_b - N_a) \ , \tag{3.100}$$

$$\frac{dN_c}{dt} = -A_{cb}\,N_c + R\,N_a \ , \tag{3.101}$$

where $S = B_{ba}\,\rho_{ba}$ is the stimulated rate. At stationarity one has

$$\frac{dN_a}{dt} = \frac{dN_b}{dt} = \frac{dN_c}{dt} = 0 \ , \tag{3.102}$$

from which we get

$$N_a = N\,\frac{A_{ba} + S}{A_{ba} + 2S + R} \ , \tag{3.103}$$

$$N_b = N\,\frac{R + S}{A_{ba} + 2S + R} \ , \tag{3.104}$$

$$N_c = \frac{R}{A_{cb}}\,N_a \ , \tag{3.105}$$

and the population imbalance $\zeta = (N_a - N_b)/N$ reads

$$\zeta = \frac{A_{ba} - R}{A_{ba} + 2S + R} .$$ (3.106)

This equation clearly shows that for $R > A_{ba}$ one gets $\zeta < 0$, thus one obtains the desired population imbalance.

3.5   Life-Time and Natural Line-Width

We have seen that the Einstein coefficient $A_{ba} = W_{ba}^{spont}$ gives the transition probability per unit of time from the atomic state $|b\rangle$ to the atomic state $|a\rangle$. This means that, according to Einstein, in the absence of an external electromagnetic radiation one has

$$\frac{dN_b}{dt} = -A_{ba} \, N_b$$ (3.107)

with the unique solution

$$N_b(t) = N_b(0) \, e^{-A_{ba}t} .$$ (3.108)

It is then quite natural to consider $1/A_{ba}$ as the characteristic time of this spontaneous transition.

More generally, the life-time τ_b of an atomic state $|b\rangle$ is defined as the reciprocal of the total spontaneous transition probability per unit time to all possible final atomic states $|a\rangle$, namely

$$\tau_b = \frac{1}{\sum_a A_{ba}} .$$ (3.109)

Clearly, if $|b\rangle$ is the ground-state then $A_{ba} = 0$ and $\tau_b = \infty$.

On the basis of the time-energy indetermination principle of Werner Heisenberg, in the radiation energy spectrum the natural line-width Γ_N due to the transition from the state $|b\rangle$ to the state $|a\rangle$ can be defined as

$$\Gamma_N = \hbar \left(\frac{1}{\tau_b} + \frac{1}{\tau_a} \right) .$$ (3.110)

Indeed it is possible to prove that in this transition the intensity of the emitted electromagnetic radiation follows the Lorentzian peak

$$I(\epsilon) = \frac{I_0 \Gamma_N^2/4}{(\epsilon - E_{ba})^2 + \Gamma_N^2/4} ,$$ (3.111)

where $\epsilon = \hbar\omega$ is the energy of the emitted photon and $E_{ba} = E_b - E_a$ is the energy difference of the two atomic states. The Lorentzian peak is centered on $\epsilon = E_{ba}$ and Γ_N is its full width at half-maximum.

It is important to observe that the effective line-width Γ measured in the experiments is usually larger than Γ_N because the radiating atoms move and collide. In fact, one can write

$$\Gamma = \Gamma_N + \Gamma_C + \Gamma_D , \tag{3.112}$$

where in addition to the natural width Γ_N there are the so-called collisional broadening width Γ_C and the Doppler broadening width Γ_D.

3.5.1 Collisional Broadening

The collisional broadening is due to the collision among the radiating atoms of a gas. The collision reduces the effective life-time of an atomic state and the collisional width can be then written as

$$\Gamma_C = \frac{\hbar}{\tau_{col}} , \tag{3.113}$$

where τ_{col} is the collisional time, i.e. the average time between two collision of atoms in the gas. According to the results of statistical mechanics, τ_{col} is given by

$$\tau_{col} = \frac{1}{n\sigma v_{mp}} , \tag{3.114}$$

where n is the number density of atoms, σ is the cross-section, and v_{mp} is the most probable speed of the particles in the gas.

In the quantum theory of scattering one usually assumes that an incoming particle, described by a plane wave along the z-axis, scatters with a target, modelled by an external potential $V(\mathbf{r})$, such that the outcoming asymptotic wavefunction of the particle reads

$$\psi(\mathbf{r}) = e^{i\mathbf{k}\cdot\mathbf{r}} + f_k(\theta)\frac{e^{ikr}}{r} , \tag{3.115}$$

where $f_k(\theta)$ is the scattering amplitude. The differential cross section is given by $d\sigma/d\Omega = |f_k(\theta)|^2$, which is the scattering probability per unit time of the particle with mass m, wavevector $\mathbf{k}$ and energy $\hbar^2 k^2/(2m)$. If one considers only s-wave scattering the differential cross section does not depend on the angle θ, and the total scattering cross section is just

$$\sigma(k) = 4\pi|f_k|^2 = \frac{4\pi}{k^2}\sin^2(\delta_0(k)) , \tag{3.116}$$

where $\delta_0(k)$ is the so-called s-wave phase shift of the cross-section. The s-wave scattering length a_s, which characterizes the effective size of the target, is then defined as the following low-energy limit

$$a_s = -\lim_{k \to 0} \frac{1}{k} \tan(\delta_0(k)) . \tag{3.117}$$

In this way one clearly finds

$$\lim_{k \to 0} \sigma(k) = 4\pi a_s^2 . \tag{3.118}$$

Thus, for a dilute and cold gas of atoms the cross-section σ of low-energy atoms can be indeed approximated as

$$\sigma = 4\pi a_s^2 , \tag{3.119}$$

where the value of a_s can be indeed determined by solving the reduced single-particle Schrödinger equation with the actual inter-atomic potential $V(\mathbf{r})$. Moreover, by considering the Maxwell-Boltzmann distribution of speeds in an ideal gas, the most probable speed v_{mp} is given by

$$v_{mp} = \sqrt{\frac{2k_B T}{m}} , \tag{3.120}$$

where T is the absolute temperature, k_B is the Boltzmann constant and m is the mass of each particle.

3.5.2 Doppler Broadening

The Doppler broadening is due to the Doppler effect caused by the distribution of velocities of atoms. For non-relativistic velocities ($v_x \ll c$) the Doppler shift in frequency is

$$\omega = \omega_0 \left(1 - \frac{v_x}{c}\right) , \tag{3.121}$$

where ω is the observed angular frequency, ω_0 is the rest angular frequency, v_x is the component of the atom speed along the axis between the observer and the atom and c is the speed of light. The Maxwell-Boltzmann distribution $f(v_x)$ of speeds $v_x = -c(\omega - \omega_0)/\omega_0$ at temperature T, given by

$$f(v_x) \, dv_x = \left(\frac{m}{2\pi k_B T}\right)^{1/2} e^{-mv_x^2/(2k_B T)} \, dv_x , \tag{3.122}$$

becomes

$$f(\omega)\, d\omega = \left(\frac{m}{2\pi k_B T}\right)^{1/2} e^{-mc^2(\omega-\omega_0)^2/(2\omega_0^2 k_B T)} \frac{c}{\omega_0}\, d\omega \qquad (3.123)$$

in terms of the angular frequency ω. This is the distribution of frequencies seen by the observer, and the full width at half-maximum of the Gaussian is taken as Doppler width, namely

$$\Gamma_D = \sqrt{\frac{8\ln(2)k_B T}{mc^2}}\, \hbar\omega_0 . \qquad (3.124)$$

3.6 Minimal Coupling and Center of Mass

Up to now the interaction between atom and electromagnetic radiation has been investigated by taking into account only the electronic degrees of freedom of the atoms. Here we consider also the effect of the atomic nucleus and for simplicity we study the hydrogen atom, whose classical complete Hamiltonian reads

$$H_{matt} = \frac{\mathbf{p}_p^2}{2m_p} + \frac{\mathbf{p}_e^2}{2m_e} - \frac{e^2}{4\pi\varepsilon_0|\mathbf{r}_p - \mathbf{r}_e|} , \qquad (3.125)$$

where $\mathbf{r}_p$ and $\mathbf{r}_e$ are the positions of proton and electron, and $\mathbf{p}_p$ and $\mathbf{p}_e$ are their corresponding linear momenta, with m_p the proton mass and m_e the electron mass.

By imposing the minimal coupling to the electromagnetic vector potential $\mathbf{A}(\mathbf{r})$, in the Coulomb gauge we get

$$H_{shift} = \frac{(\mathbf{p}_p - e\mathbf{A}(\mathbf{r}_p))^2}{2m_p} + \frac{(\mathbf{p}_e + e\mathbf{A}(\mathbf{r}_e))^2}{2m_e} - \frac{e^2}{4\pi\varepsilon_0|\mathbf{r}_p - \mathbf{r}_e|} . \qquad (3.126)$$

We now introduce center of mass position and relative position vectors:

$$\mathbf{r}_{cm} = \frac{m_p\mathbf{r}_p + m_e\mathbf{r}_e}{M} , \qquad (3.127)$$

$$\mathbf{r}_{rel} = \mathbf{r}_e - \mathbf{r}_p , \qquad (3.128)$$

where $M = m_p + m_e$ is the total mass and $\mu = m_p m_e/M$ is the reduced mass, and also the center of mass momentum and relative momentum vectors:

$$\mathbf{p}_{cm} = \mathbf{p}_p + \mathbf{p}_e , \qquad (3.129)$$

$$\mathbf{p}_{rel} = \frac{m_p}{M}\mathbf{p}_e - \frac{m_e}{M}\mathbf{p}_p . \qquad (3.130)$$

First of all we notice that

$$H_{matt} = \frac{\mathbf{p}_{cm}^2}{2M} + \frac{\mathbf{p}_{rel}^2}{2\mu} - \frac{e^2}{4\pi\varepsilon_0|\mathbf{r}_{rel}|} ,$$ (3.131)

moreover

$$H_{shift} = H_{matt} + H_I ,$$ (3.132)

where

$$\begin{aligned}
H_I ={}& -\frac{e}{M}\mathbf{p}_{cm} \cdot \mathbf{A}(\mathbf{r}_{cm} - \frac{m_e}{M}\mathbf{r}_{rel}) + \frac{e}{M}\mathbf{p}_{cm} \cdot \mathbf{A}(\mathbf{r}_{cm} + \frac{m_p}{M}\mathbf{r}_{rel}) \\
& + \frac{e}{m_p}\mathbf{p}_{rel} \cdot \mathbf{A}(\mathbf{r}_{cm} - \frac{m_e}{M}\mathbf{r}_{rel}) + \frac{e}{m_e}\mathbf{p}_{rel} \cdot \mathbf{A}(\mathbf{r}_{cm} + \frac{m_p}{M}\mathbf{r}_{rel}) \\
& + \frac{e^2}{2m_p}\mathbf{A}(\mathbf{r}_{cm} - \frac{m_e}{M}\mathbf{r}_{rel})^2 + \frac{e^2}{2m_e}\mathbf{A}(\mathbf{r}_{cm} + \frac{m_p}{M}\mathbf{r}_{rel})^2 .
\end{aligned}$$ (3.133)

In this case, the dipole approximation means

$$\mathbf{A}(\mathbf{r}_{cm} - \frac{m_e}{M}\mathbf{r}_{rel}) \simeq \mathbf{A}(\mathbf{r}_{cm} + \frac{m_p}{M}\mathbf{r}_{rel}) \simeq \mathbf{A}(\mathbf{r}_{cm})$$ (3.134)

and the interaction Hamiltonian becomes

$$H_I = \frac{e}{\mu}\mathbf{p}_{rel} \cdot \mathbf{A}(\mathbf{r}_{cm}) + \frac{e^2}{2\mu}\mathbf{A}(\mathbf{r}_{cm})^2 .$$ (3.135)

We stress that, within the dipole approximation, in the Hamiltonian H_I the dependence on $\mathbf{p}_{cm}$ disappears and that the only dependence on the center of mass is included in $\mathbf{A}(\mathbf{r}_{cm})$. Indeed, this Hamiltonian is precisely the dipolar Hamiltonian of Eq. (3.20) by using the center of mass as origin of the coordinate system.

3.7   Solved Problems

Problem 3.1

10^8 sodium atoms are excited in the first excited state of sodium by absorption of light. Knowing that the excitation energy is 2.125 eV and the lifetime is 16 ns, calculate the maximum of the emitting power.

Solution

The total absorbed energy is given by

$$E = N\epsilon ,$$

where N is the number of atoms (and also the number of absorbed photons) while ϵ is the transition energy

$$\epsilon = 2.125 \text{ eV} = 2.125 \times 1.6 \times 10^{-19} \text{ J} = 3.4 \times 10^{-19} \text{ J} .$$

The absorbed energy is then

$$E = 10^8 \times 3.4 \times 10^{-19} \text{ J} = 3.4 \times 10^{-11} \text{ J} .$$

The absorbed energy is equal to the energy emitted by spontaneous de-excitation. The emitting power $P(t)$ decays exponentially with time t as

$$P(t) = P_0 e^{-t/\tau} ,$$

where P_0 is the maximum of the emitting power and $\tau = 16$ ns is the lifetime of the excited state. It must be

$$E = \int_0^\infty P(t) \, dt = P_0 \, \tau ,$$

from which we get

$$P_0 = \frac{E}{\tau} = \frac{3.4 \times 10^{-11} \text{ J}}{16 \times 10^{-9} \text{ s}} = 2.1 \times 10^{-3} \frac{\text{J}}{\text{s}} = 2.1 \text{ mW} .$$

Problem 3.2
Derive the selection rule $\Delta m = 0$ of the matrix element $\langle n'l'm'|\hat{z}|nlm\rangle$ of the hydrogen atoms by using the formula $[\hat{L}_z, \hat{z}] = 0$.

Solution
Remember that for any eigenstate $|nlm\rangle$ of the electron in the hydrogen atom one has

$$\hat{L}_z|nlm\rangle = \hbar m \, |nlm\rangle .$$

Moreover

$$[\hat{L}_z, \hat{z}] = \hat{L}_z\hat{z} - \hat{z}\hat{L}_z .$$

We observe that

$$\begin{aligned}
\langle n'l'm'|[\hat{L}_z, \hat{z}]|nlm\rangle &= \langle n'l'm'|\hat{L}_z\hat{z} - \hat{z}\hat{L}_z|nlm\rangle \\
&= \langle n'l'm'|\hat{L}_z\hat{z}|nlm\rangle - \langle n'l'm'|\hat{z}\hat{L}_z|nlm\rangle \\
&= m'\hbar \, \langle n'l'm'|\hat{z}|nlm\rangle - \langle n'l'm'|\hat{z}|nlm\rangle \, \hbar m \\
&= (m' - m)\hbar \, \langle n'l'm'|\hat{z}|nlm\rangle .
\end{aligned}$$

Because $[\hat{L}_z, \hat{z}] = 0$ it follows that $\langle n'l'm'|[\hat{L}_z, \hat{z}]|nlm\rangle = 0$ and also

$$(m' - m)\hbar \langle n'l'm'|\hat{z}|nlm\rangle = 0 .$$

From the last expression we get

$$\langle n'l'm'|\hat{z}|nlm\rangle = 0$$

if $m' \neq m$. This is exactly the selection rule $\Delta m = 0$. We can then write

$$\langle n'l'm'|\hat{z}|nlm\rangle = \langle n'l'm|\hat{z}|nlm\rangle \, \delta_{m',m} .$$

Problem 3.3

A gas of hydrogen atoms is prepared in the state $|n = 2, l = 1, m = 0\rangle$. Calculate the electric dipole moment associated to the transition to the ground-state $|n = 1, l = 0, m = 0\rangle$.

Solution

The generic eigenfunction of the electron in the hydrogen atom in spherical coordinates is given by

$$\psi_{nlm}(r, \theta, \phi) = R_{nl}(r) \, Y_{lm}(\theta, \phi) ,$$

where $R_{nl}(r)$ is the radial wavefunction, which depends on the quantum numbers n (principal) ed l (angular momentum), while $Y_{lm}(\theta, \phi)$ is the angular wavefunction, which depends on the quantum numbers l (angular momentum) and m (projection of angular momentum).

The wavefunction of the the electron in the ground-state is

$$\psi_{100}(\mathbf{r}) = R_{10}(r)Y_{00}(\theta, \phi) ,$$

where

$$R_{10}(r) = \frac{2}{r_0^{3/2}}e^{-r/r_0}$$

$$Y_{00}(\theta, \phi) = \frac{1}{\sqrt{4\pi}} ,$$

with r_0 the Bohr radius. The wavefunction of the excited state is instead

$$\psi_{210}(\mathbf{r}) = R_{21}(r)Y_{10}(\theta, \phi) ,$$

where

$$R_{21}(r) = \frac{1}{\sqrt{3}(2r_0)^{3/2}} \frac{r}{r_0}e^{-r/(2r_0)}$$

$$Y_{10}(\theta, \phi) = \sqrt{\frac{3}{4\pi}} \cos \theta \ .$$

The vector of the dipole transition is given by

$$\mathbf{r}_T = \langle \psi_{100} | \mathbf{r} | \psi_{210} \rangle = \int d^3\mathbf{r} \ \psi_{100}^*(\mathbf{r}) \ \mathbf{r} \ \psi_{210}(\mathbf{r}) \ .$$

Remembering that in spherical coordinates

$$\mathbf{r} = (r \cos \phi \sin \theta , \ r \sin \phi \sin \theta , \ r \cos \theta)$$

with $\phi \in [0, 2\pi]$ and $\theta \in [0, \pi]$, and moreover

$$d^3\mathbf{r} = dr \ r^2 \ d\phi \ d\theta \ \sin \theta \ ,$$

the 3 components of the transition vector $\mathbf{r}_T = (x_T, y_T, z_T)$ read

$$x_T = \frac{\sqrt{3}}{4\pi} \int_0^\infty dr r^2 R_{10}(r) r R_{21}(r) \int_0^{2\pi} d\phi \cos \phi \int_0^\pi d\theta \sin^2 \theta \cos \theta \ ,$$

$$y_T = \frac{\sqrt{3}}{4\pi} \int_0^\infty dr r^2 R_{10}(r) r R_{21}(r) \int_0^{2\pi} d\phi \sin \phi \int_0^\pi d\theta \sin^2 \theta \cos \theta \ ,$$

$$z_T = \frac{\sqrt{3}}{4\pi} \int_0^\infty dr r^2 R_{10}(r) r R_{21}(r) \int_0^{2\pi} d\phi \int_0^\pi d\theta \sin \theta \cos^2 \theta \ .$$

One finds immediately that
$$x_T = y_T = 0 \ ,$$

because

$$\int_0^{2\pi} d\phi \ \cos \phi = \int_0^{2\pi} d\phi \ \sin \phi = 0 \ ,$$

while

$$\int_0^\pi d\theta \ \sin \theta \cos^2 \theta = \int_{-1}^1 d(\cos \theta) \cos^2 \theta = \frac{2}{3} .$$

As a consequence, we have

$$z_T = \frac{1}{3\sqrt{2} r_0^4} \int_0^\infty r^4 e^{-3r/(2r_0)} \ dr = \frac{2^5 r_0}{3^6 \sqrt{2}} \int_0^\infty \eta^4 e^{-\eta} \ d\eta = \frac{2^8 r_0}{3^5 \sqrt{2}} \ ,$$

because

$$\int_0^\infty \eta^n e^{-\eta}\, d\eta = \Gamma(n+1) = n! \ .$$

Finally, the modulus of the electric dipole is

$$|\mathbf{d}| = e|\mathbf{r}_T| = e\, z_T = \frac{2^8}{3^5\sqrt{2}} e\, r_0 \ ,$$

where e is the electric charge of the electron. Performing the numerical calculation we find

$$|\mathbf{d}| = 1.6 \times 10^{-19}\, \text{C} \times 0.53 \times 10^{-10}\, \text{m} \times 0.74 = 6.3 \times 10^{-30}\, \text{C m} \ .$$

Problem 3.4
Calculate the rates of spontaneous emission W^{spont} to the ground-state in a gas of hydrogen atoms which are initially:
(a) in the state $|n = 3, l = 2, m = 1\rangle$;
(b) in the state $|n = 3, l = 1, m = 1\rangle$.

Solution
The ground-state corresponds to $|n = 1, l = 0, m = 0\rangle$.
(a) In this case the transition probability is zero, within the dipolar approximation, because $\Delta l = 2$ and the permitted transitions are only with $\Delta l = \pm 1$ and $\Delta m = 0, \pm 1$.
(b) In this case the calculations must be done explicitly. The wavefunction of the electron in the ground-state is

$$\psi_{100}(\mathbf{r}) = \frac{1}{\sqrt{\pi}} \frac{1}{r_0^{3/2}} e^{-r/r_0} \ ,$$

while the wavefunction of the excited state is

$$\psi_{311}(\mathbf{r}) = \frac{2}{3^3\sqrt{\pi}} \frac{r}{r_0^{5/2}} \left(1 - \frac{r}{6r_0}\right) e^{-r/(3r_0)}\, e^{i\phi}\, \sin\theta \ ,$$

where r_0 is the Bohr radius. The vector of the dipole transition is given by

$$\mathbf{r}_T = \langle \psi_{100}|\mathbf{r}|\psi_{311}\rangle = \int d^3\mathbf{r}\, \psi_{100}^*(\mathbf{r})\, \mathbf{r}\, \psi_{311}(\mathbf{r}) \ .$$

Remembering that in spherical coordinates

$$\mathbf{r} = (r \cos\phi \sin\theta,\ r \sin\phi \sin\theta,\ r \cos\theta) \ ,$$

and also

$$d^3\mathbf{r} = dr \, r^2 \, d\phi \, d\theta \, \sin\theta \,,$$

the 3 components of the transition vector $\mathbf{r}_T = (x_T, y_T, z_T)$ read

$$x_T = -\frac{2}{3^3 \, \pi \, r_0^4} \int_0^\infty dr \, r^4 \left(1 - \frac{r}{6r_0}\right) e^{-4r/(3r_0)} \int_0^{2\pi} d\phi \, \cos\phi \, e^{i\phi} \int_0^\pi d\theta \, \sin^3\theta \,,$$

$$y_T = -\frac{2}{3^3 \, \pi \, r_0^4} \int_0^\infty dr \, r^4 \left(1 - \frac{r}{6r_0}\right) e^{-4r/(3r_0)} \int_0^{2\pi} d\phi \, \sin\phi \, e^{i\phi} \int_0^\pi d\theta \, \sin^3\theta \,,$$

$$z_T = -\frac{2}{3^3 \, \pi \, r_0^4} \int_0^\infty dr \, r^4 \left(1 - \frac{r}{6r_0}\right) e^{-4r/(3r_0)} \int_0^{2\pi} d\phi \, e^{i\phi} \int_0^\pi d\theta \, \cos\theta \, \sin^2\theta \,.$$

One find immediately that

$$z_T = 0 \,,$$

because

$$\int_0^\pi d\theta \, \cos\theta \, \sin^2\theta = 0 \,,$$

while

$$\int_0^\pi d\theta \, \sin^3\theta = \frac{2^2}{3} \,.$$

In addition, we get

$$\int_0^{2\pi} d\phi \, \cos\phi \, e^{i\phi} = \int_0^{2\pi} d\phi \, \cos^2\phi = \pi \,,$$

and similarly

$$\int_0^{2\pi} d\phi \, \sin\phi \, e^{i\phi} = i \int_0^{2\pi} d\phi \, \sin^2\phi = i \, \pi \,.$$

Instead, for the radial part

$$\int_0^\infty dr \, r^4 \left(1 - \frac{r}{6r_0}\right) e^{-4r/(3r_0)} = \frac{3^7}{2^{10}} r_0^5 \,,$$

and consequently

$$x_T = \left(-\frac{2}{3^3 \, \pi \, r_0^4}\right) \left(\frac{3^7}{2^{10}} r_0^5\right) \left(\pi \frac{2^2}{3}\right) = -\frac{3^3}{2^7} r_0$$

$$y_T = \left(-\frac{2}{3^3 \pi r_0^4}\right)\left(\frac{3^7}{2^{10}} r_0^5\right)\left(i \pi \frac{2^2}{3}\right) = -i \frac{3^3}{2^7} r_0 .$$

The squared modulus of the transition vector is given by

$$|\mathbf{r}_T|^2 = |x_T|^2 + |y_T|^2 = 2 \frac{3^6}{2^{14}} r_0^2 = 0.088 \, r_0^2 = 2.48 \times 10^{-22} \text{ m}^2 ,$$

where $r_0 = 5.29 \times 10^{-11}$ m.
The rate of spontaneous transition is given by the formula

$$W^{spont} = \frac{4}{3c^2} \, \alpha \, \omega_T^3 \, |\mathbf{r}_T|^2 ,$$

where $c = 3 \times 10^8$ m/s, $\alpha = e^2/(4\pi\epsilon_0\hbar c) \simeq 1/137$,

$$\omega_T = \frac{E_3 - E_1}{\hbar} = \frac{13.60 \text{ eV}}{\hbar}\left(1 - \frac{1}{3^2}\right) = \frac{13.60 \times 1.60 \times 10^{-19} \text{J}}{1.05 \times 10^{-34} \text{ J s/rad}} \frac{8}{9} = 1.84 \times 10^{16} \text{ rad/s} .$$

Finally, the rate of spontaneous transition reads

$$W^{spont} = 1.67 \times 10^8 \text{ rad/s} .$$

Problem 3.5

A gas of hydrogen atoms is irradiated from all directions by light with spectral density

$$\rho(\omega) = \rho_0 \, \exp\left(-\frac{(\omega - \omega_0)^2}{1000 \, \omega_0^2}\right) ,$$

where ω_0 is the frequency of the α-Lymann (1s–2p) transition.
(a) Calculate the value of ρ_0 which produces stimulated emission 2p → 1s with the rate $W^{stimul} = 12.5 \times 10^8$ s^{-1}.
(b) Determine the effective electric field E_{rms} of the incoming light, defined as

$$\frac{1}{2}\epsilon_0 E_{rms}^2 = \int_0^\infty \rho(\omega) \, d\omega .$$

Solution

(a) From the Einstein relations we know that the rate of stimulated emission W^{stimul} can be written as

$$W^{stimul} = \frac{\pi}{3\epsilon_0\hbar^2}|\mathbf{d}|^2\rho(\omega_0) ,$$

where **d** is the electric dipole momentum. In this transition we are considering

$$\mathbf{d} = -e\langle 2p|\mathbf{r}|1s\rangle ,$$

where e is the electric charge of the electron. This dipole has been previously calculated and its modulus is equal to

$$|\mathbf{d}| = 6.3 \times 10^{-30} \text{ C m} .$$

Because in this case $\rho(\omega_0) = \rho_0$, we obtain

$$\rho_0 = \frac{3\epsilon_0 \hbar^2}{\pi} \frac{W^{stimul}}{|\mathbf{d}|^2} .$$

Finally, we get

$$\rho_0 = 2.95 \times 10^{-12} \text{ Js/m}^3 .$$

(b) The integral to be calculated is

$$\mathcal{E} = \int_0^\infty \rho(\omega) \, d\omega = \int_0^\infty \rho_0 \, \exp\left(-\frac{(\omega - \omega_0)^2}{1000 \, \omega_0^2}\right) d\omega .$$

With the position

$$t = \frac{\omega - \omega_0}{\sqrt{1000} \, \omega_0}$$

the integral becomes

$$\mathcal{E} = \sqrt{1000}\rho_0 \, \omega_0 \int_{-\frac{1}{\sqrt{1000}}}^\infty \exp\left(-t^2\right) dt \simeq 10\sqrt{10} \, \rho_0 \, \omega_0 \int_0^\infty \exp\left(-t^2\right) dt$$

$$= 10\sqrt{10} \, \rho_0 \, \omega_0 \frac{\sqrt{\pi}}{2} = 5\sqrt{10\pi} \, \rho_0 \, \omega_0 .$$

The frequency ω_0 is

$$\omega_0 = -\frac{13.6 \, \text{eV}}{\hbar} \left(\frac{1}{2^2} - \frac{1}{1^2}\right) = 1.55 \times 10^{16} \, \text{s}^{-1} ,$$

and consequently

$$\mathcal{E} = 1.28 \times 10^6 \text{ J/m}^3 .$$

The effective electric field reads

$$E_{rms} = \sqrt{\frac{2\mathcal{E}}{\epsilon_0}} = 5.38 \times 10^8 \text{ V/m} ,$$

where $\epsilon_0 = 8.85 \times 10^{-12}$ J/(mV^2) is the dielectric constant in the vacuum.

Problem 3.6
Calculate both natural and Doppler width of the α-Lymann line in a hydrogen gas at the temperature 1000 K, knowing that the lifetime of the excited state is 0.16×10^{-8} s and the wavelength of the transition is 1214×10^{-10} m.

Solution
The α-Lymann line is associated to the $1s \rightarrow 2p$ transition of the hydrogen atom. The natural width between the states $|i\rangle$ and $|j\rangle$ is given by

$$\Delta \nu_N = \frac{1}{2\pi} \left(\frac{1}{\tau_i} + \frac{1}{\tau_j} \right)$$

where τ_i is the lifetime of the state $|i\rangle$ and τ_j is the lifetime of the state $|j\rangle$. In our problem $\tau_{1s} = \infty$, because $|1s\rangle$ is the ground-state, while $\tau_{2p} = 0.16 \times 10^{-8}$ s for the excited state $|2p\rangle$. It follows that

$$\Delta \nu_N = \frac{1}{2\pi} \left(\frac{1}{\tau_{1s}} + \frac{1}{\tau_{1p}} \right) = \frac{1}{6.28} \left(0 + \frac{1}{0.16 \times 10^{-8}} \right) s^{-1} = 9.9 \times 10^7 \text{ Hz} .$$

The Doppler width depends instead on the temperature $T = 1000$ °K and on the frequency ν of the transition according to the formula

$$\Delta \nu_D = \nu \sqrt{8 \ln(2)} \sqrt{\frac{k_B T}{m_H c^2}} = \frac{\nu}{c} \sqrt{8 \ln(2)} \sqrt{\frac{k_B T}{m_H}} ,$$

where $k_B = 1.3 \times 10^{-23}$ J/K is the Boltzmann constant, $m_H = 1.6 \times 10^{-27}$ kg is the mass of an hydrogen atom, while $c = 3 \times 10^8$ m/s is the speed of light in the vacuum. We know that

$$\frac{\nu}{c} = \frac{1}{\lambda} = \frac{1}{1.216 \times 10^{-7} \text{ m}} = 8.22 \times 10^6 \text{ m}^{-1} ,$$

and then we get

$$\Delta \nu_D = 8.22 \times 10^6 \times 2.35 \times \sqrt{\frac{1.3 \times 10^{-23} \times 10^3}{1.6 \times 10^{-27}}} \text{ Hz} = 5.6 \times 10^{10} \text{ Hz} .$$

Problem 3.7

Calculate the pressure (collisional) width of the α-Lymann line for a gas of hydrogen atoms with density 10^{12} atoms/m^3 and collisional cross-section 10^{-19} m^2 at the temperature 10^3 K.

Solution

The collisional width is given by

$$\Delta \nu_C = \frac{1}{2\pi \tau_{col}} \, ,$$

where τ_{col} is the collision time. This time depends on the cross-section $\sigma = 10^{-19}$ m^2 and on the gas density $n = 10^{12}$ m^{-3} according to the formula

$$\tau_{col} = \frac{1}{n \, \sigma \, v_{mp}} \, ,$$

where $v_{mp} = \sqrt{2k_B T / m_H}$ is the velocity corresponding the the maximum of the Maxwell-Boltzmann distribution. We have then

$$\Delta \nu_C = \frac{n \, \sigma}{2\pi} \sqrt{\frac{2k_B T}{m_H}} \, .$$

Because $T = 10^3$ K, $k_B = 1.3 \times 10^{-23}$ J/K and $m_H = 1.6 \times 10^{-27}$ kg, we finally obtain

$$\Delta \nu_C = 6.4 \times 10^{-5} \text{ Hz} \, .$$

Problem 3.8

Helium atoms in a gas absorb light in the transition $a \to b$, where $|a\rangle$ and $|b\rangle$ are two excited states. Knowing that the wavelength of the transition is 501.7 nm and that the lifetimes are $\tau_b = 1.4$ ns and $\tau_a = 1$ ms, calculate: (a) the natural width; (b) the Doppler width. The gas is at the temperature 1000 K.

Solution

The natural width of the spectral line between the states $|a\rangle$ and $|b\rangle$ is given by

$$\Delta \nu_N = \frac{1}{2\pi} \left(\frac{1}{\tau_a} + \frac{1}{\tau_b} \right) \, .$$

Consequently we obtain

$$\Delta \nu_N = \frac{1}{6.28} \left(\frac{1}{10^{-3}} + \frac{1}{1.4 \times 10^{-9}} \right) \text{ Hz} = 1.14 \times 10^8 \text{ Hz} \, .$$

The Doppler width is instead

$$\Delta\nu_D = \frac{\nu}{c}\sqrt{8\ln(2)}\sqrt{\frac{k_B T}{m_{He}}} \;,$$

where $T = 10^3$ K is the temperature of the gas, $k_B = 1.3 \times 10^{-23}$ J/K is the Boltzmann constant, $m_{He} = 4 \times 1.6 \times 10^{-27}$ kg is the mass of a helium atom, while $c = 3 \times 10^8$ m/s is the velocity of light in the vacuum. We know that

$$\frac{\nu}{c} = \frac{1}{\lambda} = \frac{1}{501.7 \times 10^{-9}\text{ m}} = 1.99 \times 10^6 \text{ m}^{-1} \;,$$

and then we get

$$\Delta\nu_D = 6.77 \times 10^9 \text{ Hz} \;.$$

Further Reading

For classical and quantum electrodynamics and radiative transitions:
F. Mandl and G. Shaw, Quantum Field Theory, Chap. 1, Sects. 1.3 and 1.4 (Wiley, New York, 1984).
For lifetime and line widths:
B.H. Bransden and C.J. Joachain: Physics of Atoms and Molecules, Chap. 4, Sects. 4.6 and 4.7 (Prentice Hall, Upper Saddle River, 2003).

Chapter 4
The Spin of the Electron

In this chapter we explain the origin of the intrinsic angular momentum, also known as the spin, of the electron on the basis on the Dirac equation. After introducing the Dirac equation for a relativistic massive and charged particle coupled to the electromagnetic field, we study its non relativistic limit deriving the Pauli equation. This Pauli equation, which is nothing but the Schrödinger equation with an additional term containing the spin operator, predicts very accurately the magnetic moment of the electron. Finally, we discuss the relativistic hydrogen atom and the fine-structure corrections to the non relativistic one.

4.1 The Dirac Equation

The classical energy of a nonrelativistic free particle is given by

$$E = \frac{\mathbf{p}^2}{2m} , \tag{4.1}$$

where $\mathbf{p}$ is the linear momentum and m the mass of the particle. The Schrödinger equation of the corresponding quantum particle with wavefunction $\psi(\mathbf{r}, t)$ is easily obtained by imposing the quantization prescription

$$E \to i\hbar\frac{\partial}{\partial t} , \qquad \mathbf{p} \to -i\hbar\nabla . \tag{4.2}$$

In this way one gets the time-dependent Schrödinger equation of the free particle, namely

$$i\hbar\frac{\partial}{\partial t}\psi(\mathbf{r}, t) = -\frac{\hbar^2}{2m}\nabla^2\psi(\mathbf{r}, t) , \tag{4.3}$$

© Springer International Publishing AG 2017
L. Salasnich, *Quantum Physics of Light and Matter*, UNITEXT for Physics,
DOI 10.1007/978-3-319-52998-1_4

obtained for the first time in 1926 by Erwin Schrödinger. The classical energy of a relativistic free particle is instead given by

$$E = \sqrt{\mathbf{p}^2 c^2 + m^2 c^4} \, , \tag{4.4}$$

where c is the speed of light in the vacuum. By applying directly the quantization prescription (4.2) one finds

$$i\hbar \frac{\partial}{\partial t} \psi(\mathbf{r}, t) = \sqrt{-\hbar^2 c^2 \nabla^2 + m^2 c^4} \, \psi(\mathbf{r}, t) \, . \tag{4.5}$$

This equation is quite suggestive but the square-root operator on the right side is a very difficult mathematical object. For this reason in 1927 Oskar Klein and Walter Gordon suggested to start with

$$E^2 = \mathbf{p}^2 c^2 + m^2 c^4 \tag{4.6}$$

and then to apply the quantization prescription (4.2). In this way one obtains

$$-\hbar^2 \frac{\partial^2}{\partial t^2} \psi(\mathbf{r}, t) = \left(-\hbar^2 c^2 \nabla^2 + m^2 c^4 \right) \psi(\mathbf{r}, t) \tag{4.7}$$

the so-called Klein-Gordon equation, which can be re-written in the following form

$$\left(\frac{1}{c^2} \frac{\partial^2}{\partial t^2} - \nabla^2 + \frac{m^2 c^2}{\hbar^2} \right) \psi(\mathbf{r}, t) = 0 \, , \tag{4.8}$$

i.e. a generalization of Maxwell's wave equation for massive particles. This equation has two problems: (i) it admits solutions with negative energy; (ii) the space integral over the entire space of the non negative probability density $\rho(\mathbf{r}, t) = |\psi(\mathbf{r}, t)|^2 \geq 0$ is generally not time-independent, namely

$$\frac{d}{dt} \int_{\mathbb{R}^3} \rho(\mathbf{r}, t) \, d^3 \mathbf{r} \neq 0 \, . \tag{4.9}$$

Nowadays we know that to solve completely these two problems it is necessary to promote $\psi(\mathbf{r}, t)$ to a quantum field operator. Within this second-quantization (quantum field theory) approach the Klein-Gordon equation is now used to describe relativistic particles with spin zero, like the pions or the Higgs boson.

In 1928 Paul Dirac proposed a different approach to the quantization of the relativistic particle. To solve the problem of Eq. (4.9) he considered a wave equation with only first derivatives with respect to time and space and introduced the classical energy

$$E = c \, \hat{\boldsymbol{\alpha}} \cdot \mathbf{p} + \hat{\beta} mc^2 \, , \tag{4.10}$$

such that squaring it one recovers Eq. (4.6). This condition is fulfilled only if $\hat{\alpha} = (\hat{\alpha}_1, \hat{\alpha}_2, \hat{\alpha}_3)$ and $\hat{\beta}$ satisfy the following algebra of matrices

$$\hat{\alpha}_1^2 = \hat{\alpha}_2^2 = \hat{\alpha}_3^2 = \hat{\beta}^2 = \hat{I} \,, \tag{4.11}$$

$$\hat{\alpha}_i \hat{\alpha}_j + \hat{\alpha}_j \hat{\alpha}_i = \hat{0} \,, \qquad i \neq j \tag{4.12}$$

$$\hat{\alpha}_i \hat{\beta} + \hat{\beta} \hat{\alpha}_i = \hat{0} \,, \qquad \forall i \tag{4.13}$$

where $\hat{I}$ is the identity matrix and $\hat{0}$ is the zero matrix. The smallest dimension in which the so-called Dirac matrices $\hat{\alpha}_i$ and $\hat{\beta}$ can be realized is four. In particular, one can write

$$\hat{\alpha}_i = \begin{pmatrix} \hat{0}_2 & \hat{\sigma}_i \\ \hat{\sigma}_i & \hat{0}_2 \end{pmatrix} \,, \qquad \hat{\beta} = \begin{pmatrix} \hat{I}_2 & \hat{0}_2 \\ \hat{0}_2 & -\hat{I}_2 \end{pmatrix} \,, \tag{4.14}$$

where $\hat{I}_2$ is the 2×2 identity matrix, $\hat{0}_2$ is the 2×2 zero matrix, and

$$\hat{\sigma}_1 = \begin{pmatrix} 0 & 1 \\ 1 & 0 \end{pmatrix} \qquad \hat{\sigma}_2 = \begin{pmatrix} 0 & -i \\ i & 0 \end{pmatrix} \qquad \hat{\sigma}_3 = \begin{pmatrix} 1 & 0 \\ 0 & -1 \end{pmatrix} \tag{4.15}$$

are the Pauli matrices. Equation (4.10) with the quantization prescription (4.2) gives

$$i\hbar \frac{\partial}{\partial t} \Psi(\mathbf{r}, t) = \left(-i\hbar c \, \hat{\alpha} \cdot \nabla + \hat{\beta} mc^2 \right) \Psi(\mathbf{r}, t) \,, \tag{4.16}$$

which is the Dirac equation for a free particle. Notice that the wavefunction $\Psi(\mathbf{r}, t)$ has four components in the abstract space of Dirac matrices, i.e. this spinor field can be written

$$\Psi(\mathbf{r}, t) = \begin{pmatrix} \psi_1(\mathbf{r}, t) \\ \psi_2(\mathbf{r}, t) \\ \psi_3(\mathbf{r}, t) \\ \psi_4(\mathbf{r}, t) \end{pmatrix} \,. \tag{4.17}$$

In explicit matrix form the Dirac equation is thus given by

$$i\hbar \frac{\partial}{\partial t} \begin{pmatrix} \psi_1(\mathbf{r}, t) \\ \psi_2(\mathbf{r}, t) \\ \psi_3(\mathbf{r}, t) \\ \psi_4(\mathbf{r}, t) \end{pmatrix} = \hat{H} \begin{pmatrix} \psi_1(\mathbf{r}, t) \\ \psi_2(\mathbf{r}, t) \\ \psi_3(\mathbf{r}, t) \\ \psi_4(\mathbf{r}, t) \end{pmatrix} \tag{4.18}$$

where

$$\hat{H} = \begin{pmatrix} mc^2 & 0 & -i\hbar c \frac{\partial}{\partial z} & -i\hbar c(\frac{\partial}{\partial x} - i\frac{\partial}{\partial y}) \\ 0 & mc^2 & -i\hbar c(\frac{\partial}{\partial x} + i\frac{\partial}{\partial y}) & i\hbar c \frac{\partial}{\partial z} \\ -i\hbar c \frac{\partial}{\partial z} & -i\hbar c(\frac{\partial}{\partial x} - i\frac{\partial}{\partial y}) & -mc^2 & 0 \\ -i\hbar c(\frac{\partial}{\partial x} + i\frac{\partial}{\partial y}) & i\hbar c \frac{\partial}{\partial z} & 0 & -mc^2 \end{pmatrix} .$$

$$(4.19)$$

It is easy to show that the Dirac equation satisfies the differential law of current conservation. In fact, left-multiplying Eq. (4.16) by

$$\Psi^+(\mathbf{r}, t) = \left(\psi_1^*(\mathbf{r}, t), \psi_2^*(\mathbf{r}, t), \psi_3^*(\mathbf{r}, t), \psi_4^*(\mathbf{r}, t) \right) \tag{4.20}$$

we get

$$i\hbar\Psi^+ \frac{\partial\Psi}{\partial t} = -i\hbar c \, \Psi^+ \hat{\boldsymbol{\alpha}} \cdot \nabla\Psi + mc^2 \Psi^+ \hat{\beta} \, \Psi \, . \tag{4.21}$$

Considering the Hermitian conjugate of the Dirac equation (4.16) and right-multiplying it by $\Psi(\mathbf{r}, t)$ we find instead

$$-i\hbar \frac{\partial\Psi^+}{\partial t} \Psi = i\hbar c \, \hat{\boldsymbol{\alpha}} \cdot \nabla\Psi^+ \Psi + mc^2 \Psi^+ \hat{\beta} \, \Psi \, . \tag{4.22}$$

Subtracting the last two equations we obtain the continuity equation

$$\frac{\partial}{\partial t} \rho(\mathbf{r}, t) + \nabla \cdot \mathbf{j}(\mathbf{r}, t) = 0 \, , \tag{4.23}$$

where

$$\rho(\mathbf{r}, t) = \Psi^+(\mathbf{r}, t)\Psi(\mathbf{r}, \mathbf{t}) = \sum_{i=1}^{4} |\psi_i(\mathbf{r}, t)|^2 \tag{4.24}$$

is the probability density, and $\mathbf{j}(\mathbf{r}, t)$ is the probability current with three components

$$j_k(\mathbf{r}, t) = c \, \Psi^+(\mathbf{r}, t)\hat{\alpha}_k \Psi(\mathbf{r}, t) \, . \tag{4.25}$$

Finally, we observe that from the continuity equation (4.23) one finds

$$\frac{d}{dt} \int_{\mathbb{R}^3} \rho(\mathbf{r}, t) \, d^3\mathbf{r} = 0 \, , \tag{4.26}$$

by using the divergence theorem and imposing a vanishing current density on the border at infinity. Thus, contrary to the Klein-Gordon equation, the Dirac equation does not have the probability density problem.

4.2   The Pauli Equation and the Spin

In this section we analyze the non-relativistic limit of the Dirac equation. Let us suppose that the relativistic particle has the electric charge q. In presence of an electromagnetic field, by using the Gauge-invariant substitution

$$i\hbar\frac{\partial}{\partial t} \to i\hbar\frac{\partial}{\partial t} - q\,\phi(\mathbf{r}, t) \tag{4.27}$$

$$-i\hbar\mathbf{\nabla} \to -i\hbar\mathbf{\nabla} - q\mathbf{A}(\mathbf{r}, t) \tag{4.28}$$

in Eq. (4.16), we obtain

$$i\hbar\frac{\partial}{\partial t}\Psi(\mathbf{r}, t) = \left(c\,\hat{\boldsymbol{\alpha}} \cdot (\hat{\mathbf{p}} - q\mathbf{A}(\mathbf{r}, t)) + \hat{\beta}\,mc^2 + q\,\phi(\mathbf{r}, t)\right)\Psi(\mathbf{r}, t)\,, \tag{4.29}$$

where $\hat{\mathbf{p}} = -i\hbar\mathbf{\nabla}$, $\phi(\mathbf{r}, t)$ is the scalar potential and $\mathbf{A}(\mathbf{r}, t)$ the vector potential.

To workout the non-relativistic limit of Eq. (4.29) it is useful to set

$$\Psi(\mathbf{r}, t) = e^{-imc^2 t/\hbar}\begin{pmatrix} \psi_1(\mathbf{r}, t) \\ \psi_2(\mathbf{r}, t) \\ \chi_1(\mathbf{r}, t) \\ \chi_2(\mathbf{r}, t) \end{pmatrix} = e^{-imc^2 t/\hbar}\begin{pmatrix} \psi(\mathbf{r}, t) \\ \chi(\mathbf{r}, t) \end{pmatrix}\,, \tag{4.30}$$

where $\psi(\mathbf{r}, t)$ and $\chi(\mathbf{r}, t)$ are two-component spinors, for which we obtain

$$i\hbar\frac{\partial}{\partial t}\begin{pmatrix} \psi \\ \chi \end{pmatrix} = \begin{pmatrix} q\,\phi & c\,\hat{\boldsymbol{\sigma}} \cdot (\hat{\mathbf{p}} - q\mathbf{A}) \\ c\,\hat{\boldsymbol{\sigma}} \cdot (\hat{\mathbf{p}} - q\mathbf{A}) & q\,\phi - 2mc^2 \end{pmatrix}\begin{pmatrix} \psi \\ \chi \end{pmatrix} \tag{4.31}$$

where $\hat{\boldsymbol{\sigma}} = (\hat{\sigma}_1, \hat{\sigma}_2, \hat{\sigma}_3)$. Remarkably, only in the lower equation of the system it appears the mass term mc^2, which is dominant in the non-relativistic limit. Indeed, under the approximation $\left(i\hbar\frac{\partial}{\partial t} - q\,\phi + 2mc^2\right)\chi \simeq 2mc^2\,\chi$, the previous equations become

$$\begin{pmatrix} i\hbar\frac{\partial\psi}{\partial t} \\ 0 \end{pmatrix} = \begin{pmatrix} q\,\phi & c\,\hat{\boldsymbol{\sigma}} \cdot (\hat{\mathbf{p}} - q\mathbf{A}) \\ c\,\hat{\boldsymbol{\sigma}} \cdot (\hat{\mathbf{p}} - q\mathbf{A}) & -2mc^2 \end{pmatrix}\begin{pmatrix} \psi \\ \chi \end{pmatrix}\,, \tag{4.32}$$

from which

$$\chi = \frac{\hat{\boldsymbol{\sigma}} \cdot (\hat{\mathbf{p}} - q\mathbf{A})}{2\,mc}\,\psi\,. \tag{4.33}$$

Inserting this expression in the upper equation of the system (4.32) we find

$$i\hbar\frac{\partial}{\partial t}\psi = \left(\frac{[\hat{\boldsymbol{\sigma}} \cdot (\hat{\mathbf{p}} - q\mathbf{A})]^2}{2m} + q\,\phi\right)\psi\,. \tag{4.34}$$

From the identity

$$\left[\hat{\boldsymbol{\sigma}} \cdot (\hat{\mathbf{p}} - q\mathbf{A})\right]^2 = (\hat{\mathbf{p}} - q\mathbf{A})^2 - i\,q\,(\hat{\mathbf{p}} \wedge \mathbf{A}) \cdot \hat{\boldsymbol{\sigma}} \qquad (4.35)$$

where $\hat{\mathbf{p}} = -i\hbar\mathbf{V}$, and using the relation $\mathbf{B} = \mathbf{V} \wedge \mathbf{A}$ which introduces the magnetic field we finally get

$$i\hbar\frac{\partial}{\partial t}\psi(\mathbf{r}, t) = \left(\frac{(-i\hbar\mathbf{V} - q\mathbf{A}(\mathbf{r}, t))^2}{2m} - \frac{q}{m}\mathbf{B}(\mathbf{r}, t) \cdot \hat{\mathbf{S}} + q\,\phi(\mathbf{r}, t)\right)\psi(\mathbf{r}, t) , \qquad (4.36)$$

that is the so-called Pauli equation with

$$\hat{\mathbf{S}} = \frac{\hbar}{2}\hat{\boldsymbol{\sigma}} . \qquad (4.37)$$

the spin operator. This equation was introduced in 1927 (a year before the Dirac equation) by Wolfgang Pauli as an extension of the Schrödinger equation with the phenomenological inclusion of the spin operator. If the magnetic field $\mathbf{B}$ is constant, the vector potential can be written as

$$\mathbf{A} = \frac{1}{2}\mathbf{B} \wedge \mathbf{r} \qquad (4.38)$$

and then

$$\left(\hat{\mathbf{p}} - q\mathbf{A}\right)^2 = \hat{\mathbf{p}}^2 - 2q\mathbf{A} \cdot \hat{\mathbf{p}} + q^2\mathbf{A}^2 = \hat{\mathbf{p}}^2 - q\mathbf{B} \cdot \hat{\mathbf{L}} + q^2(\mathbf{B} \wedge \mathbf{r})^2 , \qquad (4.39)$$

with $\hat{\mathbf{L}} = \mathbf{r} \wedge \hat{\mathbf{p}}$ the orbital angular momentum operator. Thus, the Pauli equation for a particle of charge q in a constant magnetic field reads

$$i\hbar\frac{\partial}{\partial t}\psi(\mathbf{r}, t) = \left(-\frac{\hbar^2\mathbf{V}^2}{2m} - \frac{q}{2m}\mathbf{B} \cdot \left(\hat{\mathbf{L}} + 2\hat{\mathbf{S}}\right) + \frac{q^2}{2m}(\mathbf{B} \wedge \mathbf{r})^2 + q\,\phi(\mathbf{r}, t)\right)\psi(\mathbf{r}, t) . \qquad (4.40)$$

In conclusion, we have shown that the spin $\hat{\mathbf{S}}$ naturally emerges from the Dirac equation. Moreover, the Dirac equation predicts very accurately the magnetic moment $\boldsymbol{\mu}_S$ of the electron ($q = -e$, $m = m_e$) which appears in the spin energy $E_s = -\hat{\boldsymbol{\mu}}_S \cdot \mathbf{B}$ where

$$\boldsymbol{\mu}_S = -g_e\frac{\mu_B}{\hbar}\hat{\mathbf{S}} \qquad (4.41)$$

with gyromagnetic ratio $g_e = 2$ and Bohr magneton $\mu_B = e\hbar/(2m) \simeq 5.79 \times 10^{-5}$ eV/T.

4.3 Dirac Equation with a Central Potential

We now consider the stationary Dirac equation with the confining spherically-symmetric potential $V(r) = V(|\mathbf{r}|)$, namely

$$\left(-i\hbar c\, \hat{\boldsymbol{\alpha}} \cdot \nabla + \hat{\beta}\, mc^2 + V(r)\right) \Phi(\mathbf{r}) = E\, \Phi(\mathbf{r}) \,. \tag{4.42}$$

This equation is easily derived from Eq. (4.29) setting $\mathbf{A} = \mathbf{0}$, $q\phi = V(r)$, and

$$\Psi(\mathbf{r}, t) = e^{-iEt/\hbar}\, \Phi(\mathbf{r}) \,. \tag{4.43}$$

The relativistic Hamiltonian

$$\hat{H} = -i\hbar c\, \hat{\boldsymbol{\alpha}} \cdot \nabla + \hat{\beta}\, mc^2 + V(r) \tag{4.44}$$

commutes with the total angular momentum operator

$$\hat{\mathbf{J}} = \hat{\mathbf{L}} + \hat{\mathbf{S}} = \mathbf{r} \wedge \hat{\mathbf{p}} + \frac{\hbar}{2}\hat{\boldsymbol{\sigma}} \tag{4.45}$$

because the external potential is spherically symmetric. In fact, one can show that

$$[\hat{H}, \hat{\mathbf{L}}] = -i\hbar c\, \hat{\boldsymbol{\alpha}} \wedge \hat{\mathbf{p}} = -[\hat{H}, \hat{\mathbf{S}}] \,. \tag{4.46}$$

Consequently one has

$$[\hat{H}, \hat{\mathbf{J}}] = \mathbf{0} \,, \tag{4.47}$$

and also

$$[\hat{H}, \hat{J}^2] = 0 \,, \qquad [\hat{J}^2, \hat{J}_x] = [\hat{J}^2, \hat{J}_y] = [\hat{J}^2, \hat{J}_z] = 0 \,, \tag{4.48}$$

where the three components $\hat{J}_x, \hat{J}_y, \hat{J}_z$ of the total angular momentum $\hat{\mathbf{J}} = (\hat{J}_x, \hat{J}_y, \hat{J}_z)$ satisfy the familiar commutation relations

$$[\hat{J}_i, \hat{J}_j] = i\hbar\, \epsilon_{ijk}\, \hat{J}_k \tag{4.49}$$

with

$$\epsilon_{ijk} = \begin{cases} 1 & \text{if } (i, j, k) \text{ is } (x, y, z) \text{ or } (z, x, y) \text{ or } (y, z, x) \\ -1 & \text{if } (i, j, k) \text{ is } (x, z, y) \text{ or } (z, y, x) \text{ or } (y, x, z) \\ 0 & \text{if } i = j \text{ or } i = k \text{ or } j = k \end{cases} \tag{4.50}$$

the Levi-Civita symbol (also called Ricci-Curbastro symbol). Note that these commutation relations can be symbolically synthesized as

$$\hat{\mathbf{J}} \wedge \hat{\mathbf{J}} = i\hbar\, \hat{\mathbf{J}} \,. \tag{4.51}$$

Indicating the states which are simultaneous eigenstates of $\hat{H}$, $\hat{J}^2$ and $\hat{J}_z$ as $|njm_j\rangle$, one has

$$\hat{H}|njm_j\rangle = E_{nj}|njm_j\rangle \,, \tag{4.52}$$

$$\hat{J}^2|njm_j\rangle = \hbar^2 j(j+1)|njm_j\rangle \,, \tag{4.53}$$

$$\hat{J}_z|njm_j\rangle = \hbar m_j|njm_j\rangle \,, \tag{4.54}$$

where j is the quantum number of the total angular momentum and $m_j = -j, -j + 1, -j+2, ..., j-2, j-1, j$ the quantum number of the third component of the total angular momentum.

In conclusion, we have found that, contrary to the total angular momentum $\hat{\mathbf{J}}$, the orbital angular momentum $\hat{\mathbf{L}}$ and the spin $\hat{\mathbf{S}}$ are not constants of motion of a particle in a central potential.

4.3.1 Relativistic Hydrogen Atom and Fine Splitting

Let us consider now the electron of the hydrogen atom. We set $q = -e$, $m = m_e$ and

$$V(r) = -\frac{e^2}{4\pi\varepsilon_0|\mathbf{r}|} = -\frac{e^2}{4\pi\varepsilon_0 r} \,. \tag{4.55}$$

Then the eigenvalues E_{nj} of $\hat{H}$ are given by

$$E_{nj} = \frac{mc^2}{\sqrt{1 + \dfrac{\alpha^2}{\left(n-j-\frac{1}{2}+\sqrt{(j+\frac{1}{2})^2-\alpha^2}\right)^2}}} - mc^2 \,, \tag{4.56}$$

with $\alpha = e^2/(4\pi\epsilon_0\hbar c) \simeq 1/137$ the fine-structure constant. We do not prove this remarkable quantization formula, obtained independently in 1928 by Charles Galton Darwin and Walter Gordon, but we stress that expanding it in powers of the fine-structure constant α to order α^4 one gets

$$E_{nj} = E_n^{(0)}\left[1 + \frac{\alpha^2}{n}\left(\frac{1}{j+\frac{1}{2}} - \frac{3}{4n}\right)\right] \,, \tag{4.57}$$

where

$$E_n^{(0)} = -\frac{1}{2}mc^2\frac{\alpha^2}{n^2} = -\frac{13.6 \, eV}{n^2} \tag{4.58}$$

Fig. 4.1 Fine splitting for the hydrogen atom. On the *left* there are the non-relativistic energy levels obtained by solving the Schrödinger equation with a Coulomb potential (Coulomb). On the *right* there are the energy levels obtained by taking into account relativistic corrections (Fine Structure)

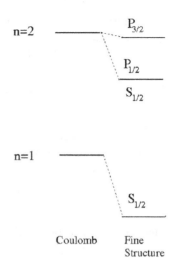

is the familiar Bohr quantization formula of the non relativistic hydrogen atom. The term which corrects the Bohr formula, given by

$$\Delta E = E_n^{(0)} \frac{\alpha^2}{n} \left(\frac{1}{j + \frac{1}{2}} - \frac{3}{4n} \right) , \tag{4.59}$$

is called fine splitting correction. This term removes the non relativistic degeneracy of energy levels, but not completely: double-degenerate levels remain with the same quantum numbers n and j but different orbital quantum number $l = j \pm 1/2$.

We have seen that, strictly speaking, in the relativistic hydrogen atom nor the orbital angular momentum $\hat{\mathbf{L}}$ nor the spin $\hat{\mathbf{S}}$ are constants of motion. As a consequence l, m_l, s and m_s are not good quantum numbers. Nevertheless, in practice, due to the smallness of fine-splitting corrections, one often assumes without problems that both $\hat{\mathbf{L}}$ and $\hat{\mathbf{S}}$ are approximately constants of motion (Fig. 4.1).

4.3.2 Relativistic Corrections to the Schrödinger Hamiltonian

It is important to stress that the relativistic Hamiltonian $\hat{H}$ of the Dirac equation in a spherically-symmetric potential, given by Eq. (4.44), can be expressed as the familiar non relativistic Schrödinger Hamiltonian

$$\hat{H}_0 = -\frac{\hbar^2}{2m} \nabla^2 + V(r) \tag{4.60}$$

plus an infinite sum of relativistic quantum corrections. To this aim one can start from the Dirac equation, written in terms of bi-spinors, i.e. Eq. (4.31) with $\mathbf{A}(\mathbf{r}, t) = \mathbf{0}$ and $q\phi(\mathbf{r}, t) = V(r)$, which gives

$$E \begin{pmatrix} \tilde{\psi} \\ \tilde{\chi} \end{pmatrix} = \begin{pmatrix} V(r) & c\,\hat{\boldsymbol{\sigma}} \cdot \hat{\mathbf{p}} \\ c\,\hat{\boldsymbol{\sigma}} \cdot \hat{\mathbf{p}} & V(r) - 2mc^2 \end{pmatrix} \begin{pmatrix} \tilde{\psi} \\ \tilde{\chi} \end{pmatrix} \tag{4.61}$$

setting $\psi(\mathbf{r}, t) = \tilde{\psi}(\mathbf{r})\, e^{-iEt/\hbar}$ and $\chi(\mathbf{r}, t) = \tilde{\chi}(\mathbf{r})\, e^{-iEt/\hbar}$. The lower equation of this system can be written as

$$\tilde{\chi} = \frac{c\,\hat{\boldsymbol{\sigma}} \cdot \hat{\mathbf{p}}}{E - V(r) + 2mc^2}\, \tilde{\psi}. \tag{4.62}$$

This is an exact equation. If $E - V(r) \ll 2mc^2$ the equation becomes

$$\tilde{\chi} = \frac{\hat{\boldsymbol{\sigma}} \cdot \hat{\mathbf{p}}}{2mc}\, \tilde{\psi}, \tag{4.63}$$

which is exactly the stationary version of Eq. (4.33) with $\mathbf{A}(\mathbf{r}, t) = \mathbf{0}$. We can do something better by expanding Eq. (4.62) with respect to the small term $(E - V(r))/2mc^2$ obtaining

$$\tilde{\chi} = \frac{\hat{\boldsymbol{\sigma}} \cdot \hat{\mathbf{p}}}{2mc}\left(1 - \frac{E - V(r)}{2mc^2} + \ldots\right)\tilde{\psi}. \tag{4.64}$$

Inserting this expression in the upper equation of the system and neglecting the higher order terms symbolized by the three dots, and after some tedious calculations, one finds

$$E\,\tilde{\psi} = \hat{H}\,\tilde{\psi}, \tag{4.65}$$

where

$$\hat{H} = \hat{H}_0 + \hat{H}_1 + \hat{H}_2 + \hat{H}_3, \tag{4.66}$$

with

$$\hat{H}_1 = -\frac{\hbar^4}{8m^3 c^2}\nabla^4, \tag{4.67}$$

$$\hat{H}_2 = \frac{1}{2m^2 c^2}\frac{1}{r}\frac{dV(r)}{dr}\mathbf{L} \cdot \mathbf{S}, \tag{4.68}$$

$$\hat{H}_3 = \frac{\hbar^2}{8m^2 c^2}\nabla^2 V(r), \tag{4.69}$$

with $\hat{H}_1$ the relativistic correction to the electron kinetic energy, $\hat{H}_2$ the spin-orbit correction, and $\hat{H}_3$ the Darwin correction.

If the external potential V(r) is that of the hydrogen atom, i.e. $V(r) = -e^2/(4\pi\varepsilon_0|\mathbf{r}|)$, one finds immediately that $H_3 = (\hbar^2 e^2)/(8m^2c^2\varepsilon_0)\delta(\mathbf{r})$ because $\nabla^2(1/|\mathbf{r}|) = -4\pi\delta(\mathbf{r})$. In addition, by applying the first-order perturbation theory to $\hat{H}$ with $\hat{H}_0$ unperturbed Hamiltonian, one gets exactly Eq. (4.57) of fine-structure correction. Physically one can say that the relativistic fine structure is due to the coupling between the spin $\hat{\mathbf{S}}$ and the orbital angular momentum $\hat{\mathbf{L}}$ of the electron. Moreover, we observe that

$$\hat{\mathbf{L}} \cdot \hat{\mathbf{S}} = \frac{1}{2}\left(\hat{J}^2 - \hat{L}^2 - \hat{S}^2\right) \tag{4.70}$$

because

$$\hat{J}^2 = (\hat{\mathbf{L}} + \hat{\mathbf{S}})^2 = \hat{L}^2 + \hat{S}^2 + 2\,\hat{\mathbf{L}} \cdot \hat{\mathbf{S}}\,, \tag{4.71}$$

and this means that the Hamiltonian $\hat{H}$ of Eq. (4.66) commutes with $\hat{L}^2$ and $\hat{S}^2$ but $\hat{H}$ does not commute with $\hat{L}_z$ and $\hat{S}_z$.

Actually, also the nucleus (the proton in the case of the hydrogen atom) has its spin $\hat{\mathbf{I}}$ which couples to electronic spin to produce the so-called hyperfine structure. However, typically, hyperfine structure has energy shifts orders of magnitude smaller than the fine structure.

4.4 Solved Problems

Problem 4.1
One electron is set in a uniform magnetic field $\mathbf{B} = (0, 0, B_0)$. Calculate the expectation value of the spin $\hat{\mathbf{S}}$ along the x axis if at $t = 0$ the spin is along the z axis.

Solution
The Hamiltonian operator of the spin is

$$\hat{H} = -\hat{\boldsymbol{\mu}}_S \cdot \mathbf{B}\,,$$

where $\hat{\mu}$ is the magnetic dipole moment of the electron, given by

$$\boldsymbol{\mu}_S = -g_e \frac{\mu_B}{\hbar}\,\hat{\mathbf{S}} = -\frac{1}{2}\,g_e\,\mu_B\,\hat{\boldsymbol{\sigma}}\,,$$

with $g_e = 2.002319 \simeq 2$ the gyromagnetic ratio of the electron, $\mu_B = e\hbar/(2m) = 9.27 \times 10^{-24}\ J/T$ the Bohr magneton and $\hat{\boldsymbol{\sigma}} = (\hat{\sigma}_1, \hat{\sigma}_2, \hat{\sigma}_3)$ the vector of Pauli matrices:

$$\hat{\sigma}_1 = \begin{pmatrix} 0 & 1 \\ 1 & 0 \end{pmatrix}$$

$$\hat{\sigma}_2 = \begin{pmatrix} 0 & -i \\ i & 0 \end{pmatrix}$$

$$\hat{\sigma}_3 = \begin{pmatrix} 1 & 0 \\ 0 & -1 \end{pmatrix} \, .$$

The Hamiltonian operator can be written as

$$\hat{H} = \frac{1}{2} \hbar \omega_0 \, \hat{\sigma}_3 = \frac{1}{2} \hbar \omega_0 \begin{pmatrix} 1 & 0 \\ 0 & -1 \end{pmatrix} \, ,$$

where $\omega_0 = g_e \mu_B B_0 / \hbar$ is the Larmor frequency of the system, with $\hbar = h/(2\pi)$ the reduced Planck constant. The initial state of the system is

$$|\psi(0)\rangle = \begin{pmatrix} 1 \\ 0 \end{pmatrix} = |\uparrow\rangle \, ,$$

while

$$\begin{pmatrix} 0 \\ 1 \end{pmatrix} = |\downarrow\rangle$$

is the state along the third component of spin. The state at time t is then given by

$$|\psi(t)\rangle = e^{-i\hat{H}t/\hbar} |\psi(0)\rangle = e^{-i\omega_0 \hat{\sigma}_3 t/2} |\uparrow\rangle = e^{-i\omega_0 t/2} |\uparrow\rangle \, ,$$

because

$$\hat{\sigma}_3 |\uparrow\rangle = |\uparrow\rangle$$

and

$$e^{-i\omega_0 \hat{\sigma}_3 t/2} |\uparrow\rangle = e^{-i\omega_0 t/2} |\uparrow\rangle \, .$$

The expectation value at time t of the spin component along the x axis is then

$$\langle \hat{S}_x(t) \rangle = \langle \psi(t)| \frac{\hbar}{2} \hat{\sigma}_1 |\psi(t)\rangle = \frac{\hbar}{2} \langle \uparrow | e^{i\omega_0 t/2} \hat{\sigma}_1 e^{-i\omega_0 t/2} |\uparrow\rangle = \frac{\hbar}{2} \langle \uparrow | \hat{\sigma}_1 |\uparrow\rangle \, .$$

Observing the

$$\hat{\sigma}_1 |\uparrow\rangle = |\downarrow\rangle \, ,$$

and also

$$\langle \uparrow | \downarrow \rangle = 0 \, ,$$

we conclude that

$$\langle \hat{S}_x(t) \rangle = 0 \, .$$

This means that if initially the spin is in the same direction of the magnetic field it remains in that direction forever: it is a stationary state with a time-dependence only in the phase. Instead, the components of spin which are orthogonal to the magnetic field have always zero expectation value.

Problem 4.2

One electron is set in a uniform magnetic field $\mathbf{B} = (0, 0, B_0)$. Calculate the expectation value of the spin $\hat{\mathbf{S}}$ along the x axis if at $t = 0$ the spin is along the x axis.

Solution

The Hamiltonian operator of spin is given by

$$\hat{H} = \frac{1}{2} \hbar \omega_0 \, \hat{\sigma}_3$$

where $\omega_0 = g_e \mu_B B_0 / \hbar$ is the Larmor frequency of the system, with g_e gyromagnetic factor and μ_B Bohr magneton. The initial state of the system is

$$|\psi(0)\rangle = \frac{1}{\sqrt{2}} (|\uparrow\rangle + |\downarrow\rangle) \,,$$

because

$$\hat{\sigma}_1 |\psi(0)\rangle = \frac{1}{\sqrt{2}} (\hat{\sigma}_1 |\uparrow\rangle + \sigma_1 |\downarrow\rangle) = \frac{1}{\sqrt{2}} (|\downarrow\rangle + |\uparrow\rangle) = |\psi(0)\rangle \,.$$

The state at time t is then

$$|\psi(t)\rangle = e^{-i\hat{H}t/\hbar} |\psi(0)\rangle = e^{-i\omega_0 \hat{\sigma}_3 t/2} \frac{1}{\sqrt{2}} (|\uparrow\rangle + |\downarrow\rangle) = \frac{1}{\sqrt{2}} \left(e^{-i\omega_0 t/2} |\uparrow\rangle + e^{i\omega_0 t/2} |\downarrow\rangle \right) \,,$$

because

$$\hat{\sigma}_3 |\uparrow\rangle = |\uparrow\rangle$$

$$\hat{\sigma}_3 |\downarrow\rangle = -|\downarrow\rangle$$

and

$$e^{-i\omega_0 \hat{\sigma}_3 t/2} |\uparrow\rangle = e^{-i\omega_0 t/2} |\uparrow\rangle$$

$$e^{-i\omega_0 \hat{\sigma}_3 t/2} |\downarrow\rangle = e^{i\omega_0 t/2} |\downarrow\rangle \,.$$

The expectation value at time t of the spin component along x axis reads

$$\langle \hat{S}_x(t) \rangle = \langle \psi(t) | \frac{\hbar}{2} \hat{\sigma}_1 | \psi(t) \rangle = \frac{\hbar}{4} \left(\langle\uparrow| e^{i\omega_0 t/2} + \langle\downarrow| e^{-i\omega_0 t/2} \right) \hat{\sigma}_1 \left(e^{-i\omega_0 t/2} |\uparrow\rangle + e^{i\omega_0 t/2} |\downarrow\rangle \right)$$

$$= \frac{\hbar}{4} \left(e^{i\omega_0 t} \langle\uparrow | \uparrow\rangle + e^{-i\omega_0 t} \langle\downarrow | \downarrow\rangle \right) = \frac{\hbar}{2} \cos(\omega_0 t) \,.$$

Problem 4.3
One electron with spin $\hat{\mathbf{S}}$, initially along z axis, is under the action of a time-dependent magnetic field $\mathbf{B} = B_1(\cos(\omega t)\mathbf{e}_x + \sin(\omega t)\mathbf{e}_y)$. Calculate the spin state at time t.

Solution
The Hamiltonian operator of spin is given by

$$\hat{H}(t) = \frac{1}{2}\hbar\omega_1 \left(\cos(\omega t)\,\hat{\sigma}_1 + \sin(\omega t)\,\hat{\sigma}_2\right) ,$$

where $\hbar\omega_1 = g_e\mu_B B_1$, with g_e gyromagnetic factor of the electron and μ_B Bohr magneton. In matrix form the Hamiltonian operator reads

$$\hat{H}(t) = \frac{\hbar}{2} \begin{pmatrix} 0 & \omega_1 e^{-i\omega t} \\ \omega_1 e^{i\omega t} & 0 \end{pmatrix} ,$$

while the spin initial state is

$$|\psi(0)\rangle = |\uparrow\rangle = \begin{pmatrix} 1 \\ 0 \end{pmatrix} .$$

The spin state at time t can be written instead as

$$|\psi(t)\rangle = a(t)|\uparrow\rangle + b(t)|\downarrow\rangle = a(t)\begin{pmatrix} 1 \\ 0 \end{pmatrix} + b(t)\begin{pmatrix} 0 \\ 1 \end{pmatrix} ,$$

where $a(t)$ and $b(t)$ are time-dependent functions to be determined with initial conditions

$$a(0) = 1 , \qquad b(0) = 0 .$$

The time-dependent Schrödinger equation

$$i\hbar\frac{\partial}{\partial t}|\psi(t)\rangle = H(t)|\psi(t)\rangle ,$$

can be rewritten as

$$i\hbar\begin{pmatrix} \dot{a}(t) \\ \dot{b}(t) \end{pmatrix} = \frac{\hbar}{2}\begin{pmatrix} 0 & \omega_1 e^{-i\omega t} \\ \omega_1 e^{i\omega t} & 0 \end{pmatrix}\begin{pmatrix} a(t) \\ b(t) \end{pmatrix} .$$

Thus the complex functions $a(t)$ and $b(t)$ satisfy the following system of first-order ordinary differential equations:

$$2i\,\dot{a}(t) = \omega_1 e^{-i\omega t}\, b(t) ,$$

$$2i\,\dot{b}(t) = \omega_1 e^{i\omega t}\, a(t) .$$

The time-dependent phase factors can be removed introducing the new variables $c(t)$ and $d(t)$ such that

$$a(t) = e^{-i\omega t/2} c(t) ,$$

$$d(t) = e^{i\omega t/2} d(t) .$$

The differential system then becomes

$$2i\, \dot{c}(t) = -\omega\, c(t) + \omega_1\, d(t) ,$$

$$2i\, \dot{d}(t) = \omega\, d(t) + \omega_1\, c(t) .$$

This system can be solved by using the Laplace transform

$$F(s) = \int_0^\infty f(t)\, e^{-st}\, dt .$$

Indeed, by applying the Laplace transform, the system becomes

$$2i\, (s\, C(s) - c(0)) = -\omega\, C(s) + \omega_1\, D(s) .$$

$$2i\, (s\, D(s) - d(0)) = \omega\, D(s) + \omega_1\, C(s) .$$

Taking into account the initial conditions $c(0) = 1$ and $d(0) = 0$, one can get $D(s)$ from the second equation, namely

$$D(s) = \frac{\omega_1}{2is - \omega} C(s) ,$$

and insert it into the first one. In this way

$$C(s) = \frac{s}{\omega_R^2 + s^2} + i\frac{\omega}{2}\frac{1}{\omega_R^2 + s^2} ,$$

where $\omega_R = \frac{1}{2}\sqrt{\omega_1^2 + \omega^2}$ is the so-called Rabi frequency. By applying the Laplace anti-transform we immediately obtain

$$c(t) = \cos(\omega_R t) + i\frac{\omega}{2\omega_R} \sin(\omega_R t) .$$

Coming back to the initial variables $a(t)$ and $b(t)$ we eventually find

$$|a(t)|^2 = \frac{\omega^2}{4\omega_R^2} + \frac{\omega_1^2}{4\omega_R^2} \cos^2(\omega_R t) , \qquad |b(t)|^2 = \frac{\omega_1^2}{4\omega_R^2} \sin^2(\omega_R t) .$$

which clearly satisfy the relation $|a(t)|^2 + |b(t)|^2 = 1$. Notice that for $\omega = 0$ it follows $\omega_R = \omega_1/2$ from which

$$|a(t)|^2 = \cos^2 (\omega_R t) , \qquad\qquad |b(t)|^2 = \sin^2 (\omega_R t) .$$

In this case there is complete spin-flip during the dynamics.

Problem 4.4

Derive the non relativistic formula of Bohr for the spectrum of the hydrogen atom from the relativistic expression

$$E_{nj} = \frac{mc^2}{\sqrt{1 + \dfrac{\alpha^2}{\left(n-j-\frac{1}{2}+\sqrt{(j+\frac{1}{2})^2-\alpha^2}\right)^2}}} - mc^2$$

which is obtained from the Dirac equation, expanding it in powers of $\alpha \simeq 1/137$ at order α^2.

Solution

Setting $x = \alpha^2$ the relativistic spectrum can be written as

$$E(x) = mc^2 f(x) - mc^2 ,$$

where

$$f(x) = \left(1 + \frac{x}{(A + \sqrt{B - x})^2}\right)^{-1/2} ,$$

with $A = n - j - 1/2$ e $B = (j + 1/2)^2$. Let us now expand $f(x)$ in MacLaurin series at first order:

$$f(x) = f(0) + f'(0)x ,$$

where

$$f(0) = 1$$

$$f'(0) = -\frac{1}{2}\frac{1}{(A + \sqrt{B})^2} = -\frac{1}{2}\frac{1}{n^2} .$$

We get at first order in x

$$E(x) = mc^2 - \frac{1}{2}mc^2\frac{x}{n^2} - mc^2 = -\frac{1}{2}mc^2\frac{x}{n^2} ,$$

namely

$$E_n = -\frac{1}{2}mc^2\frac{\alpha^2}{n^2} ,$$

which is exactly the non relativistic Bohr formula.

Problem 4.5
Calculate the fine splitting for the state $|3p\rangle$ of the hydrogen atom.

Solution
From the relativistic formula of the energy levels of the hydrogen atom

$$E_{nj} = \frac{mc^2}{\sqrt{1 + \frac{\alpha^2}{\left(n-j-\frac{1}{2}+\sqrt{(j+\frac{1}{2})^2-\alpha^2}\right)^2}}} - mc^2$$

expanding to order α^4 we get

$$E_{nj} = E_n^{(0)} \left[1 + \frac{\alpha^2}{n}\left(\frac{1}{j+\frac{1}{2}} - \frac{3}{4n}\right)\right]$$

where

$$E_n^{(0)} = -\frac{1}{2}mc^2\frac{\alpha^2}{n^2} = -\frac{13.6\,eV}{n^2}.$$

The integer number j is the quantum number of the total angular momentum $\mathbf{J} = \mathbf{L} + \mathbf{S}$, where $j = 1/2$ if $l = 0$ and $j = l - 1/2, l + 1/2$ if $l \neq 0$. The state $|3p\rangle$ means $n = 3$ and $l = 1$, consequently $j = 1/2$ or $j = 3/2$. The hyperfine correction for $j = 1/2$ reads

$$\Delta E_{3,\frac{1}{2}} = E_3^{(0)}\frac{\alpha^2}{3}\left(1 - \frac{1}{4}\right) = E_3^{(0)}\frac{\alpha^2}{4} = -\frac{13.6\,eV}{9}\frac{1}{137^2 \cdot 4} = -2.01 \times 10^{-5}\,eV.$$

The hyperfine correction for $j = 3/2$ is instead

$$\Delta E_{3,\frac{3}{2}} = E_3^{(0)}\frac{\alpha^2}{3}\left(\frac{1}{2} - \frac{1}{4}\right) = E_3^{(0)}\frac{\alpha^2}{12} = -\frac{13.6\,eV}{9}\frac{1}{137^2 \cdot 12} = -6.71 \times 10^{-6}\,eV.$$

Further Reading

For the Dirac equation and the Pauli equation:
J.D. Bjorken and S.D. Drell, Relativistic Quantum Mechanics, Chap. 1, Sects. 1.1, 1.2, 1.3, 1.4 (McGraw-Hill, New York, 1964).
For the relativistic hydrogen atom and the fine splitting:
V.B. Berestetskii, E.M. Lifshitz, L.P. Pitaevskii, Relativistic Quantum Theory, vol. 4 of Course of Theoretical Physics, Chap. 4, Sects. 34, 35, 36 (Pergamon Press, Oxford, 1971).
B.H. Bransden and C.J. Joachain: Physics of Atoms and Molecules, Chap. 5, Sect. 5.1 (Prentice Hall, Upper Saddle River, 2003).

Chapter 5
Energy Splitting and Shift Due to External Fields

In this chapter we study the Stark effect and the Zeeman effect on atoms, and in particular on the Hydrogen atom. In the Stark effect one finds that an external constant electric field induces splitting of degenerate energy levels and energy shift of the ground-state. Similarly, in the Zeeman effect an external constant magnetic field induces splitting of degenerate energy levels, but the splitting properties strongly depend on the intensity of the external magnetic field.

5.1 Stark Effect

Let us consider the Hydrogen atom under the action of a constant electric field $\mathbf{E}$. We write the constant electric field as

$$\mathbf{E} = E\,\mathbf{u}_z = (0, 0, E), \tag{5.1}$$

choosing the z axis in the same direction of $\mathbf{E}$. The Hamiltonian operator of the system is then given by

$$\hat{H} = \hat{H}_0 + \hat{H}_I , \tag{5.2}$$

where

$$\hat{H}_0 = \frac{\hat{p}^2}{2m} - \frac{e^2}{4\pi\epsilon_0\,r} \tag{5.3}$$

is the non-relativistic Hamiltonian of the electron in the hydrogen atom (with $m = m_e$ the electron mass), while

$$\hat{H}_I = -e\,\phi(\mathbf{r}) = e\,\mathbf{E}\cdot\mathbf{r} = -\mathbf{d}\cdot\mathbf{E} = e\,E\,z \tag{5.4}$$

© Springer International Publishing AG 2017
L. Salasnich, *Quantum Physics of Light and Matter*, UNITEXT for Physics,
DOI 10.1007/978-3-319-52998-1_5

is the Hamiltonian of the interaction due to the electric scalar potential $\phi(\mathbf{r}) = -\mathbf{E} \cdot \mathbf{r}$ such that $\mathbf{E} = -\nabla \phi$, with $-e$ electric charge of the electron and $\mathbf{d} = -e\mathbf{r}$ the electric dipole.

Let $|nlm_l\rangle$ be the eigenstates of $\hat{H}_0$, such that

$$\hat{H}_0 |nlm_l\rangle = E_n^{(0)} |nlm_l\rangle \tag{5.5}$$

with

$$E_n^{(0)} = -\frac{13.6}{n^2} \text{ eV} , \tag{5.6}$$

the Bohr spectrum of the hydrogen atom, and moreover

$$\hat{L}^2 |nlm_l\rangle = \hbar^2 \, l(l+1) \, |nlm_l\rangle , \qquad \hat{L}_z |nlm_l\rangle = \hbar \, m_l \, |nlm_l\rangle . \tag{5.7}$$

At the first order of degenerate perturbation theory the energy spectrum is given by

$$E_n = E_n^{(0)} + E_n^{(1)} , \tag{5.8}$$

where $E_n^{(1)}$ is one of the eigenvalues of the submatrix $\hat{H}_I^n$, whose elements are

$$a_{l'm_l',lm_l}^{(n)} = \langle nl'm_l'|\hat{H}_I|nlm_l\rangle = e \, E \, \langle nl'm_l'|z|nlm_l\rangle . \tag{5.9}$$

Thus, in general, there is a linear splitting of degenerate energy levels due to the external electric field E. This effect is named after Johannes Stark, who discovered it in 1913. Actually, it was discovered independently in the same year also by Antonino Lo Surdo.

It is important to stress that the ground-state $|1s\rangle = |n = 1, l = 0, m_l = 0\rangle$ of the hydrogen atom is not degenerate and for it $\langle 1s|z|1s\rangle = 0$. It follows that there is no linear Stark effect for the ground-state. Thus, we need the second order perturbation theory, namely

$$E_1 = E_1^{(0)} + E_1^{(1)} + E_1^{(2)} , \tag{5.10}$$

where

$$E_1^{(1)} = \langle 100|eEz|100\rangle = 0 , \tag{5.11}$$

$$E_1^{(2)} = \sum_{nlm_l \neq 100} \frac{|\langle nlm_l|eEz|100\rangle|^2}{E_1^{(0)} - E_n^{(0)}} = e^2 E^2 \sum_{n=2}^{\infty} \frac{|\langle n10|z|100\rangle|^2}{E_1^{(0)} - E_n^{(0)}} , \tag{5.12}$$

where the last equality is due to the dipole selection rules. This formula shows that the electric field produces a quadratic shift in the energy of the ground state. This phenomenon is known as the quadratic Stark effect.

The polarizability α_p of an atom is defined in terms of the energy-shift ΔE_1 of the atomic ground state energy E_1 induced by an external electric field E as follows:

$$\Delta E_1 = -\frac{1}{2}\alpha_p E^2 .$$ (5.13)

Hence, for the hydrogen atom we can write

$$\alpha_p = -2e^2 \sum_{n=2}^{\infty} \frac{|\langle n10|z|100\rangle|^2}{E_1^{(0)} - E_n^{(0)}} = -\frac{9}{4}\frac{e^2 r_0^2}{E_1^{(0)}} ,$$ (5.14)

where the last equality is demonstrated in Problem 5.2, with r_0 the Bohr radius.

5.2 Zeeman Effect

Let us consider the hydrogen atom under the action of a constant magnetic field **B**. According to the Pauli equation, the Hamiltonian operator of the system is given by

$$\hat{H} = \frac{(\hat{\mathbf{p}} + e\mathbf{A})^2}{2m} - \frac{e^2}{4\pi\epsilon_0 r} - \hat{\mu}_S \cdot \mathbf{B}$$ (5.15)

where

$$\hat{\mu}_S = -\frac{e}{m}\hat{\mathbf{S}}$$ (5.16)

is the spin dipole magnetic moment, with $\hat{\mathbf{S}}$ the spin of the electron, and **A** is the vector potential, such that $\mathbf{B} = \nabla \wedge \mathbf{A}$. Because the magnetic field **B** is constant, the vector potential can be written as

$$\mathbf{A} = \frac{1}{2}\mathbf{B} \wedge \mathbf{r} ,$$ (5.17)

and then

$$(\hat{\mathbf{p}} + e\mathbf{A})^2 = \hat{\mathbf{p}}^2 + 2e\hat{\mathbf{p}} \cdot \mathbf{A} + e^2\mathbf{A}^2 = \hat{\mathbf{p}}^2 + 2e\mathbf{B} \cdot \hat{\mathbf{L}} + e^2(\mathbf{B} \wedge \mathbf{r})^2 ,$$ (5.18)

with $\hat{\mathbf{L}} = \mathbf{r} \wedge \hat{\mathbf{p}}$. In this way the Hamiltonian can be expressed as

$$\hat{H} = \hat{H}_0 + \hat{H}_I ,$$ (5.19)

where

$$\hat{H}_0 = \frac{\hat{p}^2}{2m} - \frac{e^2}{4\pi\epsilon_0 r} \tag{5.20}$$

is the non-relativistic Hamiltonian of the electron in the hydrogen atom, while

$$\hat{H}_I = -\hat{\boldsymbol{\mu}} \cdot \mathbf{B} + \frac{e^2}{8m}(\mathbf{B} \wedge \mathbf{r})^2 \tag{5.21}$$

is the Hamiltonian of the magnetic interaction, with

$$\hat{\boldsymbol{\mu}} = \hat{\boldsymbol{\mu}}_L + \hat{\boldsymbol{\mu}}_S = -\frac{e}{2m}(\hat{\mathbf{L}} + 2\hat{\mathbf{S}}) \tag{5.22}$$

the total dipole magnetic moment of the electron. We now write the constant magnetic field as

$$\mathbf{B} = B\,\mathbf{u}_z = (0, 0, B) \ , \tag{5.23}$$

choosing the z axis in the same direction of $\mathbf{B}$. In this way the interaction Hamiltonian becomes

$$\hat{H}_I = \frac{eB}{2m}\left(\hat{L}_z + 2\hat{S}_z\right) + \frac{e^2 B^2}{8m}(x^2 + y^2) \ . \tag{5.24}$$

The first term, called paramagnetic term, grows linearly with the magnetic field B while the second one, the diamagnetic term, grows quadratically. The paramagnetic term is of the order of $\mu_B B$, where $\mu_B = e\hbar/(2m) = 9.3 \times 10^{-24}$ J/T $= 5.29 \times 10^{-5}$ eV/T is the Bohr magneton. Because the unperturbed energy of $\hat{H}_0$ is of the order of 10 eV, the paramagnetic term can be considered a small perturbation.

5.2.1 Strong-Field Zeeman Effect

Usually the diamagnetic term is much smaller than the paramagnetic one, and becomes observable only for B of the order of $10^6/n^4$ Tesla, i.e. mainly in the astrophysical context. Thus in laboratory the diamagnetic term is usually negligible (apart for very large values of the principal quantum number n) and the effective interaction Hamiltonian reads

$$H_I = \frac{eB}{2m}\left(\hat{L}_z + 2\hat{S}_z\right) \ . \tag{5.25}$$

Thus (5.20) is the unperturbed Hamiltonian and (5.25) the perturbing Hamiltonian. It is clear that this total Hamiltonian is diagonal with respect to the eigenstates $|nlm_l m_s\rangle$ and one obtains immediately the following energy spectrum

$$E_{n,m_l,m_s} = E_n^{(0)} + \mu_B B \, (m_l + 2m_s) \, , \qquad (5.26)$$

where $E_n^{(0)}$ is the unperturbed Bohr eigenspectrum and $\mu_B = e\hbar/(2m)$ is the Bohr magneton, with m the mass of the electron. Equation (5.26) describes the high-field Zeeman effect, first observed in 1896 by Pieter Zeeman. The field B does not remove the degeneracy in l but it does remove the degeneracy in m_l and m_s. The selection rules for dipolar transitions require $\Delta m_s = 0$ and $\Delta m_l = 0, \pm 1$. Thus the spectral line corresponding to a transition $n \to n'$ is split into 3 components, called Lorentz triplet (Fig. 5.1).

Fig. 5.1 Zeeman effect for the hydrogen atom. Due to a (strong) magnetic field the energy of the degenerate excited state $|2p\rangle = |n = 2, l = 1\rangle$ is split into 3 levels and the energy of the degenerate excited state $|3d\rangle = |n = 3, l = 2\rangle$ is split into 5 levels. But the spectral lines are always three (Lorentz triplet)

5.2.2 Weak-Field Zeeman Effect

In the hydrogen atom the strong-field Zeeman effect is observable if the magnetic field B is between about $1/n^3$ Tesla and about $10^6/n^4$ Tesla, with n the principal quantum number. In fact, as previously explained, for B larger than about $10^6/n^4$ Tesla the diamagnetic term is no more negligible. Instead, for B smaller than about $1/n^3$ Tesla the splitting due to the magnetic field B becomes comparable with the splitting due to relativistic fine-structure corrections. Thus, to study the effect of a weak field B, i.e. the weak-field Zeeman effect, the unperturbed non-relativistic Hamiltonian $\hat{H}_0$ given by the Eq. (5.20) is no more reliable. One must use instead the exact relativistic Hamiltonian or, at least, the non-relativistic one with relativistic corrections, namely

$$\hat{H}_0 = \hat{H}_{0,0} + \hat{H}_{0,1} + \hat{H}_{0,2} + \hat{H}_{0,3} , \tag{5.27}$$

where

$$\hat{H}_{0,0} = -\frac{\hbar^2}{2m}\nabla^2 - \frac{e^2}{4\pi\epsilon_0 r} \tag{5.28}$$

$$\hat{H}_{0,1} = -\frac{\hbar^4}{8m^3c^2}\nabla^4 , \tag{5.29}$$

$$\hat{H}_{0,2} = \frac{1}{2m^2c^2}\frac{1}{r}\frac{dV(r)}{dr}\mathbf{L}\cdot\mathbf{S} , \tag{5.30}$$

$$\hat{H}_{0,3} = \frac{\hbar^2}{8m^2c^2}\nabla^2 V(r) , \tag{5.31}$$

with $\hat{H}_{0,0}$ the non relativistic Hamiltonian, $\hat{H}_{0,1}$ the relativistic correction to the electron kinetic energy, $\hat{H}_{0,2}$ the spin-orbit correction, and $\hat{H}_{0,3}$ the Darwin correction. In any case, m_l and m_s are no more good quantum numbers because $\hat{L}_z$ and $\hat{S}_z$ do not commute with the new $\hat{H}_0$.

The Hamiltonian (5.27) commutes instead with $\hat{L}^2$, $\hat{S}^2$, $\hat{J}^2$ and $\hat{J}_z$. Consequently, for this Hamiltonian the good quantum numbers are n, l, s, j and m_j. Applying again the first-order perturbation theory, where now (5.27) is the unperturbed Hamiltonian and (5.25) is the perturbing Hamiltonian, one obtains the following energy spectrum

$$E_{n,l,j,m_j} = E_{n,j}^{(0)} + E_{n,l,s,j,m_j}^{(1)} , \tag{5.32}$$

where $E_{n,j}^{(0)}$ is the unperturbed relativistic spectrum and

$$E_{n,l,s,j,m_j}^{(1)} = \frac{eB}{2m}\langle n, l, s, j, m_j|\hat{L}_z + 2\hat{S}_z|n, l, s, j, m_j\rangle \tag{5.33}$$

is the first-order correction, which is indeed not very easy to calculate. But we can do it. First we note that

$$\langle n, l, s, j, m_j | \hat{L}_z + 2\hat{S}_z | n, l, s, j, m_j \rangle = \langle n, l, s, j, m_j | \hat{J}_z + \hat{S}_z | n, l, s, j, m_j \rangle$$
$$= \hbar m_j + \langle n, l, s, j, m_j | \hat{S}_z | n, l, s, j, m_j \rangle \ . \tag{5.34}$$

Then, on the basis of the Wigner-Eckart theorem,[1] we have

$$\hbar^2 j(j+1)\langle n, l, s, j, m_j | \hat{S}_z | n, l, s, j, m_j \rangle = \langle n, l, s, j, m_j | (\hat{\mathbf{S}} \cdot \hat{\mathbf{J}}) \, \hat{J}_z | n, l, s, j, m_j \rangle \ , \tag{5.35}$$

from which we obtain

$$\hbar^2 j(j+1)\langle n, l, s, j, m_j | \hat{S}_z | n, l, s, j, m_j \rangle = \hbar m_j \langle n, l, s, j, m_j | \hat{\mathbf{S}} \cdot \hat{\mathbf{J}} | n, l, s, j, m_j \rangle$$
$$= \hbar m_j \langle n, l, s, j, m_j | \frac{1}{2} \left(\hat{J}^2 + \hat{S}^2 - \hat{L}^2 \right) | n, l, s, j, m_j \rangle$$
$$= \hbar m_j \frac{1}{2} \hbar^2 \left(j(j+1) + s(s+1) - l(l+1) \right) \ . \tag{5.36}$$

In conclusion, for a weak magnetic field B the first-order correction is given by

$$E^{(1)}_{n,l,s,j,m_j} = \mu_B \, B \, g_{l,s,j} \, m_j \ , \tag{5.37}$$

where $\mu_B = e\hbar/(2m)$ is the Bohr magneton, and

$$g_{l,s,j} = 1 + \frac{j(j+1) + s(s+1) - l(l+1)}{2j(j+1)} \tag{5.38}$$

is the so-called Landé factor. Clearly, in the case of the electron $s = 1/2$ and the Landé factor becomes

$$g_{l,j} = 1 + \frac{j(j+1) - l(l+1) + 3/4}{2j(j+1)} \ . \tag{5.39}$$

Strictly speaking, the energy splitting described by Eq. (5.37) is fully reliable only for a weak magnetic field B in the range 0 Tesla $\leq B \ll 1/n^3$ Tesla, with n the principal quantum number. In fact, if the magnetic field B approaches $1/n^3$ Tesla one observes a complex pattern of splitting, which moves by increasing B from the

---

[1]The Wigner-Eckart theorem states that for any vector operator $\hat{\mathbf{V}} = (\hat{V}_1, \hat{V}_2, \hat{V}_3)$ such that $[\hat{J}_i, \hat{V}_j] = i\hbar\epsilon_{ijk}\hat{V}_k$ holds the identity

$$\hbar^2 j(j+1) \langle n, l, s, j, m_j | \hat{\mathbf{V}} | n, l, s, j, m_j \rangle = \langle n, l, s, j, m_j | (\hat{\mathbf{V}} \cdot \hat{\mathbf{J}}) \hat{\mathbf{J}} | n, l, s, j, m_j \rangle \ .$$

In our case $\hat{\mathbf{V}} = \hat{\mathbf{S}}$ and we have considered the z component only.

splitting described by Eq. (5.37) towards the splitting described by Eq. (5.26). This transition, observed in 1913 by Friedrich Paschen and Ernst Back, is now called the Paschen-Back effect.

5.3 Solved Problems

Problem 5.1
Determine the Stark splitting for the state $|2p\rangle$ of the hydrogen atom in a electric field of 10^7 V/m.

Solution
We write the constant electric field as $\mathbf{E} = E\,\mathbf{u}_z$, choosing the z axis in the same direction of $\mathbf{E}$. The Hamiltonian operator is given by

$$\hat{H} = \hat{H}_0 + \hat{H}_1 \ ,$$

where

$$\hat{H}_0 = \frac{\hat{p}^2}{2m} - \frac{e^2}{4\pi\epsilon_0\,r}$$

is the Hamiltonian of the unperturbed system, while

$$\hat{H}_1 = e\,\mathbf{E}\cdot\mathbf{r} = e\,E\,z$$

is the Hamiltonian of the perturbation due to the electric field, with e electric charge of the electron. Let $|nlm_l\rangle$ be the eigenstate of $\hat{H}_0$, such that

$$\hat{H}_0|nlm_l\rangle = E_n^{(0)}|nlm_l\rangle$$

with

$$E_n^{(0)} = -\frac{13.6}{n^2}\ \mathrm{eV} \ ,$$

the Bohr spectrum of the hydrogen atom, and moreover

$$\hat{L}^2|nlm_l\rangle = \hbar\,l(l+1)\,|nlm_l\rangle \ , \qquad\qquad \hat{L}_z|nlm_l\rangle = \hbar\,m_l\,|nlm_l\rangle \ .$$

At the first order of perturbation theory the energetic spectrum is given by

$$E_n = E_n^{(0)} + E_n^{(1)} \ ,$$

where $E_n^{(1)}$ is one of the eigenvalues of the submatrix $\hat{H}_1^n$, whose elements are

$$a_{l'm'_l,lm_l}^{(n)} = \langle nl'm'_l|\hat{H}_1|nlm_l\rangle = e\,E\,\langle nl'm'_l|z|nlm_l\rangle\;.$$

If $n = 2$ then $l = 0, 1$ and $m_l = -l, -l + 1, ..., l - 1, l$, i.e. $m_l = 0$ if $l = 0$ and $m_l = -1, 0, 1$ if $l = 1$. It follows that the submatrix $\hat{H}^{(2)}$ is a 4×4 matrix, given by

$$\hat{H}_1^{(2)} = \begin{pmatrix} a_{00,00}^{(2)} & a_{00,10}^{(2)} & a_{00,11}^{(2)} & a_{00,1-1}^{(2)} \\ a_{10,00}^{(2)} & a_{10,10}^{(2)} & a_{10,11}^{(2)} & a_{10,1-1}^{(2)} \\ a_{11,00}^{(2)} & a_{11,10}^{(2)} & a_{11,11}^{(2)} & a_{11,1-1}^{(2)} \\ a_{1-1,00}^{(2)} & a_{1-1,10}^{(2)} & a_{1-1,11}^{(2)} & a_{1-1,1-1}^{(2)} \end{pmatrix}\;.$$

Note that

$$\langle nl'm'_l|z|nlm_l\rangle = \langle nl'm'_l|z|nlm_l\rangle\,\delta_{m'_l,m_l}\,\delta_{l+l',1}\;,$$

i.e. the selection rules $\Delta l = \pm 1$ e $\Delta m_l = 0$ hold. It follows that many elements of the submatrix $\hat{H}_1^{(2)}$ are zero:

$$\hat{H}_1^{(2)} = \begin{pmatrix} 0 & a_{00,10}^{(2)} & 0 & 0 \\ a_{10,00}^{(2)} & 0 & 0 & 0 \\ 0 & 0 & 0 & 0 \\ 0 & 0 & 0 & 0 \end{pmatrix}\;,$$

and also $a_{00,10}^{(2)} = a_{10,00}^{(2)}$. One finds immediately that 2 eigenvalues of $\hat{H}^{(2)}$ are zero. To determine the other 2 eigenvalues it is sufficient to study the 2×2 given by

$$\hat{H}_1^{(2)} = \begin{pmatrix} 0 & a_{00,10}^{(2)} \\ a_{00,10}^{(2)} & 0 \end{pmatrix}\;,$$

whose eigenvalues are

$$\lambda_{1,2} = \pm a_{00,10}^{(2)} = \pm e\,E\,\langle 200|z|210\rangle\;.$$

We must now calculate $\langle 200|z|210\rangle$. By using the completeness relation

$$1 = \int d^3\mathbf{r}|\mathbf{r}\rangle\langle\mathbf{r}|\;,$$

this matrix element can be written as

$$\langle 200|z|210\rangle = \langle 200|\int d^3\mathbf{r}|\mathbf{r}\rangle\langle\mathbf{r}|z|210\rangle = \int d^3\mathbf{r}\langle 200|\mathbf{r}\rangle z\langle\mathbf{r}|210\rangle = \int d^3\mathbf{r}\psi_{200}^*(\mathbf{r})z\psi_{210}^*(\mathbf{r})\;.$$

For the hydrogen atom one has

$$\psi_{200}(\mathbf{r}) = \frac{1}{2\sqrt{2\pi}r_0^{3/2}}(1 - \frac{r}{2r_0})e^{-r/(2r_0)} \ ,$$

$$\psi_{210}(\mathbf{r}) = \frac{1}{4\sqrt{2\pi}r_0^{3/2}}\frac{r}{r_0}e^{-r/(2r_0)}\cos\theta \ ,$$

where $r_0 = 0.53 \times 10^{-10}$ m is the Bohr radius. Remembering that in spherical coordinates $z = r\cos\theta$, we get

$$\int d^3\mathbf{r}\ \psi_{200}^*(\mathbf{r})z\psi_{210}(\mathbf{r}) = \int_0^\infty dr r^2 \int_0^{2\pi} d\phi \int_0^\pi d\theta \sin\theta\ \psi_{200}^*(\mathbf{r})\ r\cos\theta\ \psi_{210}^*(\mathbf{r}) =$$

$$= \frac{1}{16\pi r_0^4}\int_0^\infty dr\ e^{-r/r_0}\ r^4(1 - \frac{r}{2r_0})\int_0^{2\pi} d\phi \int_{-1}^1 d(\cos\theta)\cos^2\theta$$

$$= \frac{1}{12r_0^4}\int_0^\infty dr\ e^{-r/r_0}\ r^4(1 - \frac{r}{2r_0}) = \frac{1}{12r_0^4}(-36r_0^5) = -3r_0 \ ,$$

namely

$$\langle 200|z|210\rangle = -3r_0 \ .$$

Finally, the possible values of the perturbing energy $E_2^{(1)}$ read

$$E_2^{(1)} = \begin{cases} 3eEr_0 \\ 0 \\ -3eEr_0 \end{cases} \ .$$

It follows that the unperturbed energy $E_2^{(0)}$ splits into 3 levels, given by $E_2 = E_2^{(0)} + E_2^{(1)}$, but the central one of them coincides with the unperturbed level. In conclusion, the Stark splitting is

$$\Delta E_2 = \pm 3eEr_0 = \pm 3e \times 10^7 \frac{V}{m} \cdot 0.53 \times 10^{-10}\ m = \pm 1.6 \times 10^{-3}\ eV \ .$$

Problem 5.2
Calculate the Stark shift of the state $|1s\rangle$ of the hydrogen atom in a electric field of 10^7 V/m.

Solution

We write the constant electric field as $\mathbf{E} = E\,\mathbf{e}_z$, choosing the z axis in the same direction of $\mathbf{E}$. The Hamiltonian operator is given by

$$\hat{H} = \hat{H}_0 + \hat{H}_1 \, ,$$

where

$$\hat{H}_0 = \frac{\hat{p}^2}{2m} - \frac{e^2}{4\pi\epsilon_0 r}$$

is the Hamiltonian of the unperturbed system, while

$$\hat{H}_1 = e\,\mathbf{E}\cdot\mathbf{r} = e\,E\,z$$

is the Hamiltonian of the perturbation due to the electric field, with e electric charge of the electron. At first order of perturbation theory the energy of $|1s\rangle = |n = 1, l = 0, m_l = 0\rangle$ is given by

$$E_1 = E_1^{(0)} + E_1^{(1)} \, ,$$

where

$$E_1^{(0)} = -13.6\,\text{eV} \, ,$$

while $E_1^{(1)}$ is

$$E_1^{(1)} = \langle 100|\hat{H}_1|100\rangle = e\,E\,\langle 100|z|100\rangle = 0 \, .$$

This matrix element is zero due to the selection rules $\Delta l = \pm 1$ e $\Delta m_l = 0$. Thus, we need the second order perturbation theory, namely

$$E_1 = E_1^{(0)} + E_1^{(1)} + E_1^{(2)} \, ,$$

where

$$E_1^{(2)} = \sum_{nlm_l \neq 100} \frac{|\langle nlm_l|e\,Ez|100\rangle|^2}{E_1^{(0)} - E_n^{(0)}} = \sum_{n=2}^{\infty} \frac{|\langle n10|e\,Ez|100\rangle|^2}{E_1^{(0)} - E_n^{(0)}} \, ,$$

by using again the selection rules. Remembering that

$$E_n^{(0)} = \frac{E_1^{(0)}}{n^2} \, ,$$

we can write

$$E_1^{(2)} = \frac{e^2 E^2}{E_1^{(0)}} \sum_{n=2}^{\infty} \frac{n^2}{(n^2 - 1)} |\langle n10|z|100\rangle|^2 .$$

In addition, knowing that in spherical coordinates $z = r \cos \theta$, the matrix element $\langle n10|z|100\rangle$ reads

$$\langle n10|z|100\rangle = \int d^3 \mathbf{r} \, \psi_{n1o}^*(\mathbf{r}) z \psi_{100}(\mathbf{r}) =$$

$$= \int_0^{\infty} dr r^3 R_{n1}(r) R_{10}(r) \int_0^{2\pi} d\phi \int d\theta \sin \theta \cos \theta Y_{10}(\theta, \phi) Y_{00}(\theta, \phi) .$$

Because

$$Y_{00}(\theta, \phi) = \frac{1}{\sqrt{4\pi}} \qquad Y_{10}(\theta, \phi) = \sqrt{\frac{3}{4\pi}} \cos \theta ,$$

we get

$$\langle n10|z|100\rangle = \frac{1}{\sqrt{3}} \int_0^{\infty} dr r^3 R_{n1}(r) R_{10}(r) = \frac{1}{\sqrt{3}} r_0 f(n) ,$$

where $r_0 = 0.53 \times 10^{-10}$ m is the Bohr radius and $f(n)$ is a function of the principal quantum number n. It is possible to show that

$$\sum_{n=2}^{\infty} \frac{n^2}{n^2 - 1} f(n)^2 = \frac{27}{8} .$$

From this exact relation we obtain

$$E_1^{(2)} = \frac{e^2 E^2 r_0^2}{3 E_1^{(0)}} \frac{27}{8} = \frac{9}{8} \frac{e^2 E^2 r_0^2}{E_1^{(0)}} .$$

Inserting the numerical values we finally get

$$E_1^{(2)} = -2.3 \times 10^{-15} \text{ eV} .$$

Problem 5.3
Determine the Zeeman splitting for the states $|1s\rangle$ e $|2p\rangle$ of the hydrogen atom in a magnetic field of 10 T.

Solution

In presence of a magnetic field with intensity B between $1/n^3$ Tesla and $10^6/n^4$ Tesla, one has the strong-field Zeeman effect and the energy spectrum is given by

$$E_{n,m,m_s} = E_n + \mu_B B(m_l + 2m_s) \,,$$

where

$$E_n = -\frac{13.6 \text{ eV}}{n^2}$$

is the Bohr spectrum, m_l is the quantum number of the third component of the orbital angular momentum $\mathbf{L}$ and m_s is the quantum number of the third component of the spin $\mathbf{S}$. In addition $\mu_B = e\hbar/(2m) = 9.3 \times 10^{-24}$ J/T is the Bohr magneton. In our problem we are in the regime of validity of the strong-field Zeeman effect and

$$\mu_B B = 9.3 \times 10^{-24} \frac{\text{J}}{\text{T}} \times 10 \text{ T} = 9.3 \times 10^{-23}\text{J} = 9.3 \times 10^{-23} \times \frac{10^{19}}{1.6} \text{ eV} = 5.8 \times 10^{-4} \text{ eV} \,.$$

For the state $|1s\rangle = |n = 1, l = 0, m_l = 0\rangle$ one has $s = \frac{1}{2}$ and consequently $m_s = -\frac{1}{2}, \frac{1}{2}$. It follows that

$$\Delta E_{1,0,-\frac{1}{2}} = -\mu_B B = -5.8 \times 10^{-4} \text{ eV}$$

$$\Delta E_{1,0,\frac{1}{2}} = \mu_B B = 5.8 \times 10^{-4} \text{ eV} \,.$$

For the state $|2p\rangle = |n = 2, l = 1, m_l = -1, 0, 1\rangle$ one has $s = \frac{1}{2}$ and then $m_s = -\frac{1}{2}, \frac{1}{2}$. It follows that

$$\Delta E_{2,-1,-\frac{1}{2}} = -2\mu_B B = -11.6 \times 10^{-4} \text{ eV}$$

$$\Delta E_{2,0,-\frac{1}{2}} = -\mu_B B = -5.8 \times 10^{-4} \text{ eV}$$

$$\Delta E_{2,1,-\frac{1}{2}} = \Delta E_{2,-1,\frac{1}{2}} = 0$$

$$\Delta E_{2,0,\frac{1}{2}} = \mu_B B = 5.8 \times 10^{-4} \text{ eV}$$

$$\Delta E_{2,1,\frac{1}{2}} = 2\mu_B B = 11.6 \times 10^{-4} \text{ eV} \,.$$

Problem 5.4

Calculate the Zeeman splitting for the state $|2p\rangle$ of the hydrogen atom in a magnetic field of 5 Gauss.

Solution

The magnetic field $B = 5$ Gauss $= 5 \times 10^{-4}$ T is quite weak (the magnetic field of the Earth is about 0.5 Gauss). With a field B between 0 and 10^{-2} T, the energy spectrum is well described by

$$E_{n,l,j,m_j} = E_n + \mu_B B \, g_{l,j} \, m_j \, ,$$

where $\mu_B = e\hbar/(2m) = 9.3 \times 10^{-24}$ J/T is the Bohr magneton, m_j is the quantum number of the third component of the total angular momentum $\mathbf{J} = \mathbf{L} + \mathbf{S}$, and $g_{l,j}$ is the Landé factor, given by

$$g_{l,j} = 1 + \frac{j(j+1) - l(l+1) + 3/4}{2j(j+1)} .$$

In this problem the magnetic field is very weak, and one has the weak-field Zeeman effect. For the state $|2p\rangle = |n = 2, l = 1, m_l = -1, 0, 1\rangle$ one has $s = \frac{1}{2}$ and then $m_s = -\frac{1}{2}, \frac{1}{2}$. In addition, $j = l \pm \frac{1}{2} = \frac{1}{2}, \frac{3}{2}$ from which $m_j = -\frac{3}{2}, -\frac{1}{2}, \frac{1}{2}, \frac{3}{2}$. In the case $j = 1/2$ the Landé factor reads

$$g_{1,\frac{1}{2}} = 1 + \frac{3/4 - 2 + 3/4}{3/2} = 1 - 1/3 = \frac{2}{3} .$$

In the case $j = 3/2$ the Landè factor is instead given by

$$g_{1,\frac{3}{2}} = 1 + \frac{15/4 - 2 + 3/4}{15/2} = 1 + \frac{1}{3} = \frac{4}{3} .$$

In our problem

$$\mu_B B = 9.3 \times 10^{-24} \frac{J}{T} \times 5 \times 10^{-4} \, T = 4.6 \times 10^{-27} J = 4.6 \times 10^{-27} \times \frac{10^{19}}{1.6} \, eV = 2.9 \times 10^{-8} \, eV .$$

We can now calculate the Zeeman splittings. For $j = \frac{1}{2}$ one has $m_j = -\frac{1}{2}, \frac{1}{2}$ and then

$$\Delta E_{1,\frac{1}{2},-\frac{1}{2}} = -\frac{1}{3}\mu_B B = -0.97 \times 10^{-8} \, eV$$

$$\Delta E_{1,\frac{1}{2},\frac{1}{2}} = \frac{1}{3}\mu_B B = 0.97 \times 10^{-8} \, eV .$$

For $j = \frac{3}{2}$ one has instead $m_j = -\frac{3}{2}, -\frac{1}{2}, \frac{1}{2}, \frac{3}{2}$ and then

$$\Delta E_{1,\frac{3}{2},-\frac{3}{2}} = -2\mu_B B = -5.8 \times 10^{-8} \, eV$$

$$\Delta E_{1,\frac{3}{2},-\frac{1}{2}} = -\frac{2}{3}\mu_B B = -1.93 \times 10^{-8} \, eV$$

$$\Delta E_{1,\frac{3}{2},\frac{1}{2}} = \frac{2}{3}\mu_B B = 1.93 \times 10^{-8} \text{ eV}$$

$$\Delta E_{1,\frac{3}{2},\frac{3}{2}} = 2\mu_B B = 5.8 \times 10^{-8} \text{ eV} .$$

Further Reading

For the Stark effect:

R.W. Robinett, Quantum Mechanics: Classical Results, Modern Systems, and Visualized Examples, Chap. 19, Sect. 19.6 (Oxford Univ. Press, Oxford, 2006).

For the Zeeman effect:

B.H. Bransden and C.J. Joachain: Physics of Atoms and Molecules, Chap. 6, Sects. 6.1 and 6.2 (Prentice Hall, Upper Saddle River, 2003).

Chapter 6
Many-Body Systems

In this chapter we want to analyze atoms with many electrons and, more generally, systems with many interacting identical particles. We first consider the general properties of many identical particles with their bosonic or fermionic many-body wavefunctions, and the connection between spin and statistics which explains the Pauli principle and the main features of the periodic table of elements. We then discuss the Hartree-Fock approximation and also the density functional theory, historically introduced by Thomas and Fermi to model the electronic cloud of atoms and recently applied also to atomic Bose-Einstein condensates. Finally, we illustrate the Born-Oppenheimer approximation, which is very useful to treat molecules composed by many electrons in interaction with several nuclei.

6.1 Identical Quantum Particles

First of all, we introduce the generalized coordinate $x = (\mathbf{r}, \sigma)$ of a particle which takes into account the spatial coordinate $\mathbf{r}$ but also the intrinsic spin σ pertaining to the particle. For instance, a spin $1/2$ particle has $\sigma = -1/2, 1/2 = \downarrow, \uparrow$. By using the Dirac notation the corresponding single-particle state is

$$|x\rangle = |\mathbf{r}\, \sigma\rangle \,. \tag{6.1}$$

We now consider N identical particles; for instance particles with the same mass and electric charge. The many-body wavefunction of the system is given by

$$\Psi(x_1, x_2, ..., x_N) = \Psi(\mathbf{r}_1, \sigma_1, \mathbf{r}_2, \sigma_2, ..., \mathbf{r}_N, \sigma_N) \,, \tag{6.2}$$

© Springer International Publishing AG 2017
L. Salasnich, *Quantum Physics of Light and Matter*, UNITEXT for Physics,
DOI 10.1007/978-3-319-52998-1_6

According to quantum mechanics identical particles are indistinguishable. As a consequence, it must be

$$|\Psi(x_1, x_2, ..., x_i, ..., x_j, ..., x_N)|^2 = |\Psi(x_1, x_2, ..., x_j, ..., x_i, ..., x_N)|^2 , \qquad (6.3)$$

which means that the probability of finding the particles must be independent on the exchange of two generalized coordinates x_i and x_j. Obviously, for 2 particles this implies that

$$|\Psi(x_1, x_2)|^2 = |\Psi(x_2, x_1)|^2 . \qquad (6.4)$$

Experiments suggests that there are only two kind of identical particles which satisfy Eq. (6.3): bosons and fermions. For N identical bosons one has

$$\Psi(x_1, x_2, ..., x_i, ..., x_j, ..., x_N) = \Psi(x_1, x_2, ..., x_j, ..., x_i, ..., x_N) , \qquad (6.5)$$

i.e. the many-body wavefunction is symmetric with respect to the exchange of two coordinates x_i and x_j. Note that for 2 identical bosonic particles this implies

$$\Psi(x_1, x_2) = \Psi(x_2, x_1). \qquad (6.6)$$

For N identical fermions one has instead

$$\Psi(x_1, x_2, ..., x_i, ..., x_j, ..., x_N) = -\Psi(x_1, x_2, ..., x_j, ..., x_i, ..., x_N) , \qquad (6.7)$$

i.e. the many-body wavefunction is anti-symmetric with respect to the exchange of two coordinates x_i and x_j. Note that for 2 identical fermionic particles this implies

$$\Psi(x_1, x_2) = -\Psi(x_2, x_1) . \qquad (6.8)$$

An immediate consequence of the anti-symmetry of the fermionic many-body wavefunction is the *Pauli Principle*: if $x_i = x_j$ then the many-body wavefunction is zero. In other words: the probability of finding two fermionic particles with the same generalized coordinates is zero.

A remarkable experimental fact, which is often called *spin-statistics theorem* because can be deduced from other postulates of relativistic quantum field theory, is the following: identical particles with integer spin are bosons while identical particles with semi-integer spin are fermions. For instance, photons are bosons with spin 1 while electrons are fermions with spin $1/2$. Notice that for a composed particle it is the total spin which determines the statistics. For example, the total spin (sum of nuclear and electronic spins) of ^{4}He atom is 0 and consequently this atom is a boson, while the total spin of ^{3}He atom is $1/2$ and consequently this atom is a fermion.

6.2 Non-interacting Identical Particles

The quantum Hamiltonian of N identical non-interacting particles is given by

$$\hat{H}_0 = \sum_{i=1}^{N} \hat{h}(x_i) , \qquad (6.9)$$

where $\hat{h}(x)$ is the single-particle Hamiltonian. Usually the single-particle Hamiltonian is given by

$$\hat{h}(x) = -\frac{\hbar^2}{2m} \nabla^2 + U(\mathbf{r}) , \qquad (6.10)$$

with $U(\mathbf{r})$ the external confining potential. In general the single-particle Hamiltonian $\hat{h}$ satisfies the eigenvalue equation

$$\hat{h}(x) \, \phi_n(x) = \epsilon_n \, \phi_n(x) , \qquad (6.11)$$

where ϵ_n are the single-particle eigenenergies and $\phi_n(x)$ the single-particle eigenfunctions, with $n = 1, 2, \dots$.

The many-body wavefunction $\Psi(x_1, x_2, \dots, x_N)$ of the system can be written in terms of the single-particle wavefunctions $\phi_n(x)$ but one must take into account the spin-statistics of the identical particles. For N bosons the simplest many-body wave function reads

$$\Psi(x_1, x_2, \dots, x_N) = \phi_1(x_1) \, \phi_1(x_2) \dots \phi_1(x_N) , \qquad (6.12)$$

which corresponds to the configuration where all the particles are in the lowest-energy single-particle state $\phi_1(x)$. This is indeed a pure Bose-Einstein condensate. Note that for 2 bosons the previous expression becomes

$$\Psi(x_1, x_2) = \phi_1(x_1)\phi_1(x_2) . \qquad (6.13)$$

Obviously there are infinite configuration which satisfy the bosonic symmetry of the many-body wavefunction. For example, with 2 bosons one can have

$$\Psi(x_1, x_2) = \phi_4(x_1)\phi_4(x_2) , \qquad (6.14)$$

which means that the two bosons are both in the fourth eigenstate; another example is

$$\Psi(x_1, x_2) = \frac{1}{\sqrt{2}} (\phi_1(x_1)\phi_2(x_2) + \phi_1(x_2)\phi_2(x_1)) , \qquad (6.15)$$

where the factor $1/\sqrt{2}$ has been included to maintain the same normalization of the many-body wavefunction, and in this case the bosons are in the first two available single-particles eigenstates (Fig. 6.1).

Fig. 6.1 Ground-state of a system of identical non-interacting bosons (**a**) and fermions (**b**) is a harmonic trap

For N fermions the simplest many-body wave function is instead very different, and it is given by

$$\Psi(x_1, x_2, ..., x_N) = \frac{1}{\sqrt{N!}} \begin{pmatrix} \phi_1(x_1) & \phi_1(x_2) & ... & \phi_1(x_N) \\ \phi_2(x_1) & \phi_2(x_2) & ... & \phi_2(x_N) \\ ... & ... & ... & ... \\ \phi_N(x_1) & \phi_N(x_2) & ... & \phi_N(x_N) \end{pmatrix} \qquad (6.16)$$

that is the so-called Slater determinant of the $N \times N$ matrix obtained with the N lowest-energy single particle wavefunctions $\psi_n(x)$, with $n = 1, 2, ..., N$, calculated in the N possible generalized coordinates x_i, with $i = 1, 2, ..., N$. Note that for 2 fermions the previous expression becomes

$$\Psi(x_1, x_2) = \frac{1}{\sqrt{2}} (\phi_1(x_1)\phi_2(x_2) - \phi_1(x_2)\phi_2(x_1)) . \qquad (6.17)$$

We stress that for non-interacting identical particles the Hamiltonian (6.9) is separable and the total energy associated to the bosonic many-body wavefunction (6.12) is simply

$$E = N \epsilon_1 , \qquad (6.18)$$

while for the fermionic many-body wavefunction (6.16) the total energy (in the absence of degenerate single-particle energy levels) reads

$$E = \epsilon_1 + \epsilon_2 + ... + \epsilon_N , \qquad (6.19)$$

which is surely higher than the bosonic one. The highest occupied single-particle energy level is called Fermi energy, and it indicated as ϵ_F; in our case it is obviously $\epsilon_F = \epsilon_N$.

Notice the above definition of Fermi energy ϵ_F can be considered acceptable in many contexts and is a simple and effective definition, especially for the zero-temperature Fermi gas. Unfortunately it is imprecise in some contexts, e.g. finite systems with a nonzero gap between the last full and first empty state. A better definition, such as the limit for $T \to 0$ of the chemical potential μ (see Sect. 7.5) would lead to some intermediate value between the energy of the highest occupied state and that of the lowest empty state.

6.2.1 Uniform Gas of Non-interacting Fermions

A quite important physical system is the uniform gas of non-interacting fermions. It is indeed a good starting point for the description of all the real systems which have a finite interaction between fermions.

The non-interacting uniform Fermi gas is obtained setting to zero the confining potential, i.e.

$$U(\mathbf{r}) = 0 , \tag{6.20}$$

and imposing periodicity conditions on the single-particle wavefunctions, which are plane waves with a spinor

$$\phi(x) = \frac{1}{\sqrt{V}} e^{i\mathbf{k}\cdot\mathbf{r}} \chi_\sigma , \tag{6.21}$$

where χ_σ is the spinor for spin-up and spin-down along a chosen z asis:

$$\chi_\uparrow = \begin{pmatrix} 1 \\ 0 \end{pmatrix} , \qquad \chi_\downarrow = \begin{pmatrix} 0 \\ 1 \end{pmatrix} . \tag{6.22}$$

At the boundaries of a cube having volume V and side L one has

$$e^{ik_x(x+L)} = e^{ik_x x} , \quad e^{ik_y(y+L)} = e^{ik_y y} , \quad e^{ik_z(z+L)} = e^{ik_z z} . \tag{6.23}$$

It follows that the linear momentum $\mathbf{k}$ can only take on the values

$$k_x = \frac{2\pi}{L} n_x , \quad k_y = \frac{2\pi}{L} n_y , \quad k_z = \frac{2\pi}{L} n_z , \tag{6.24}$$

where n_x, n_y, n_z are integer quantum numbers. The single-particle energies are given by

$$\epsilon_{\mathbf{k}} = \frac{\hbar^2 k^2}{2m} = \frac{\hbar^2}{2m} \frac{4\pi^2}{L^2} (n_x^2 + n_y^2 + n_z^2) . \tag{6.25}$$

In the thermodynamic limit $L \to \infty$, the allowed values are closely spaced and one can use the continuum approximation

$$\sum_{n_x,n_y,n_z} \to \int dn_x \, dn_y \, dn_z \, , \tag{6.26}$$

which implies

$$\sum_{\mathbf{k}} \to \frac{L^3}{(2\pi)^3} \int d^3\mathbf{k} = V \int \frac{d^3\mathbf{k}}{(2\pi)^3} \, . \tag{6.27}$$

The total number N of fermionic particles is given by

$$N = \sum_{\sigma} \sum_{\mathbf{k}} \Theta \left(\epsilon_F - \epsilon_{\mathbf{k}} \right) \, , \tag{6.28}$$

where the Heaviside step function $\Theta(x)$, such that $\Theta(x) = 0$ for $x < 0$ and $\Theta(x) = 1$ for $x > 0$, takes into account the fact that fermions are occupied only up to the Fermi energy ϵ_F, which is determined by fixing N. Notice that at finite temperature T the total number N of ideal fermions is instead obtained from the Fermi-Dirac distribution, namely

$$N = \sum_{\sigma} \sum_{\mathbf{k}} \frac{1}{e^{\beta(\epsilon_{\mathbf{k}} - \mu)} + 1} \, , \tag{6.29}$$

where $\beta = 1/(k_B T)$, with k_B the Boltzmann constant, and μ is the chemical potential of the system. In the limit $\beta \to +\infty$, i.e. for $T \to 0$, the Fermi-Dirac distribution becomes the Heaviside step function and μ is identified as the Fermi energy ϵ_F.

In the continuum limit and choosing spin $1/2$ fermions one finds

$$N = \sum_{\sigma=\uparrow,\downarrow} V \int \frac{d^3\mathbf{k}}{(2\pi)^3} \Theta \left(\epsilon_F - \frac{\hbar^2 k^2}{2m} \right) \, , \tag{6.30}$$

from which one gets (the sum of spins gives simply a factor 2) the uniform density

$$\rho = \frac{N}{V} = \frac{1}{3\pi^2} \left(\frac{2m\epsilon_F}{\hbar^2} \right)^{3/2} \, . \tag{6.31}$$

The formula can be inverted giving the Fermi energy ϵ_F as a function of the density ρ, namely

$$\epsilon_F = \frac{\hbar^2}{2m} \left(3\pi^2 \rho \right)^{2/3} \, . \tag{6.32}$$

In many applications the Fermi energy ϵ_F is written as

$$\epsilon_F = \frac{\hbar^2 k_F^2}{2m} \, , \tag{6.33}$$

where k_F is the so-called Fermi wave-number, given by

$$k_F = \left(3\pi^2 \rho\right)^{1/3} \, . \tag{6.34}$$

The total energy E of the uniform and non-interacting Fermi system is given by

$$E = \sum_\sigma \sum_k \epsilon_k \, \Theta \left(\epsilon_F - \epsilon_k\right) \, , \tag{6.35}$$

and using again the continuum limit with spin $1/2$ fermions it becomes

$$E = \sum_{\sigma=\uparrow,\downarrow} V \int \frac{d^3 k}{(2\pi)^3} \frac{\hbar^2 k^2}{2m} \, \Theta \left(\epsilon_F - \frac{\hbar^2 k^2}{2m}\right) \, , \tag{6.36}$$

from which one gets the energy density

$$\mathcal{E} = \frac{E}{V} = \frac{3}{5} \rho \, \epsilon_F = \frac{3}{5} \frac{\hbar^2}{2m} \left(3\pi^2\right)^{2/3} \rho^{5/3} \tag{6.37}$$

in terms of the Fermi energy ϵ_F and the uniform density ρ.

6.2.2 Atomic Shell Structure and the Periodic Table of the Elements

The non-relativistic quantum Hamiltonian of Z identical non-interacting electrons in the neutral atom is given by

$$\hat{H}_0 = \sum_{i=1}^{Z} \hat{h}(\mathbf{r}_i) \, , \tag{6.38}$$

where $\hat{h}(\mathbf{r})$ is the single-particle Hamiltonian given by

$$\hat{h}(\mathbf{r}) = -\frac{\hbar^2}{2m} \nabla^2 + U(\mathbf{r}) \, , \tag{6.39}$$

with

$$U(\mathbf{r}) = -\frac{Ze^2}{4\pi\varepsilon_0 \, |\mathbf{r}|} \tag{6.40}$$

the confining potential due to the attractive Coulomb interaction between the single electron and the atomic nucleus of positive charge Ze, with $e > 0$.

Because the confining potential $U(\mathbf{r})$ is spherically symmetric, i.e. $U(\mathbf{r}) = U(|\mathbf{r}|)$, the single-particle Hamiltonian $\hat{h}$ satisfies the eigenvalue equation

$$\hat{h}(\mathbf{r}) \, \phi_{nlm_lm_s}(\mathbf{r}) = \epsilon_n(Z) \, \phi_{nlm_lm_s}(\mathbf{r}) \, , \tag{6.41}$$

where

$$\epsilon_n(Z) = -13.6 \, eV \, \frac{Z^2}{n^2} \tag{6.42}$$

are the Bohr single-particle eigenenergies of the hydrogen-like atom, and $\phi_{nlm_lm_s}(\mathbf{r}) = R_{nl}(r)Y_{lm_l}(\theta, \phi)$ are the single-particle eigenfunctions, which depends on the principal quantum numbers $n = 1, 2, \ldots$, the angular quantum number $l = 0, 1, \ldots, n - 1$, the third-component angular quantum number $m_l = -l, -l + 1, \ldots, l - 1, l$, and the third-component spin quantum number $m_s = -\frac{1}{2}, \frac{1}{2}$. Notice that here

$$\phi_{nlm_lm_s}(\mathbf{r}) = \phi_{nlm_l}(\mathbf{r}, \sigma) \tag{6.43}$$

with $\sigma = \uparrow$ for $m_s = \frac{1}{2}$ and $\sigma = \downarrow$ for $m_s = -\frac{1}{2}$ (Table 6.1).

Table 6.1 Lightest atoms and their ground-state energy E on the basis of single-particle energies

Z	Atom	Symbol	E
1	Hydrogen	H	$\epsilon_1(1)$
2	Helium	He	$2\epsilon_1(2)$
3	Lihium	Li	$2\epsilon_1(3) + \epsilon_2(3)$
4	Berylium	Be	$2\epsilon_1(4) + 2\epsilon_2(4)$
5	Boron	B	$2\epsilon_1(5) + 3\epsilon_2(5)$
6	Carbon	C	$2\epsilon_1(6) + 4\epsilon_2(6)$
7	Nitrogen	N	$2\epsilon_1(7) + 5\epsilon_2(7)$
8	Oxygen	O	$2\epsilon_1(8) + 6\epsilon_2(8)$
9	Fluorine	F	$2\epsilon_1(9) + 7\epsilon_2(9)$
10	Neon	Ne	$2\epsilon_1(10) + 8\epsilon_2(10)$
11	Sodium	Na	$2\epsilon_1(11) + 8\epsilon_2(11) + \epsilon_3(11)$
12	Magnesium	Mg	$2\epsilon_1(12) + 8\epsilon_2(12) + 2\epsilon_3(12)$
13	Aluminium	Al	$2\epsilon_1(13) + 8\epsilon_2(13) + 3\epsilon_3(13)$
14	Silicon	Si	$2\epsilon_1(14) + 8\epsilon_2(14) + 4\epsilon_3(14)$

Due to the Pauli principle the ground-state energy E of this system of Z electrons strongly depends of the degeneracy of single-particle energy levels. In the Table we report the ground-state energy E of the lightest atoms on the basis of their single-particle energy levels $\epsilon_n(Z)$.

The degeneracy of the single-particle energy level $\epsilon_n(Z)$ is clearly independent on Z and given by

$$deg(\epsilon_n(Z)) = \sum_{l=0}^{n-1} 2(2l+1) = 2n^2 , \qquad (6.44)$$

which is the maximum number of electrons with the same principal quantum number n. The set of states with the same principal quantum number is called theoretical shell. The number of electrons in each theoretical shell are: 2, 8, 18, 32, 52. One expects that the more stable atoms are characterized by fully occupied theoretical shells. Actually, the experimental data, namely the periodic table of elements due to Dmitri Mendeleev, suggest that the true number of electrons in each experimental shell are instead: 2, 8, 8, 18, 18, 32, because the noble atoms are characterized the following atomic numbers:

$$2, \; 2+8 = 10, \; 2+8+8 = 18, \; 2+8+8+18 = 36,$$
$$2+8+8+18+18 = 54, \; 2+8+8+18+18+32 = 86 ,$$

corresponding to Helium ($Z = 2$), Neon ($Z = 10$), Argon ($Z = 18$), Krypton ($Z = 36$), Xenon ($Z = 54$), and Radon ($Z = 86$). The experimental sequence is clearly similar but not equal to the theoretical one, due to repetitions of 8 and 18.

It is important to stress that the theoretical sequence is obtained under the very crude assumption of non-interacting electrons. To improve the agreement between theory and experiment one must include the interaction between the electrons.

6.3 Interacting Identical Particles

The quantum Hamiltonian of N identical interacting particles is given by

$$\hat{H} = \sum_{i=1}^{N} \hat{h}(x_i) + \frac{1}{2} \sum_{\substack{i,j=1 \\ i \neq j}}^{N} V(x_i, x_j) = \hat{H}_0 + \hat{H}_I , \qquad (6.45)$$

where $\hat{h}$ is the single-particle Hamiltonian and $V(x_i, x_j)$ is the inter-particle potential of the mutual interaction. In general, due to the inter-particle potential, the Hamiltonian (6.45) is not separable and the many-body wavefunctions given by Eqs. (6.12) and (6.16) are not exact eigenfunctions of $\hat{H}$.

6.3.1 Variational Principle

Many approaches to the determination of the ground-state of an interacting many-body system are based on the so-called variational principle, which is actually a theorem.

Theorem *For any normalized many-body state* $|\Psi\rangle$, *i.e. such that* $\langle\Psi|\Psi\rangle = 1$, *which belongs to the Hilbert space on which acts the Hamiltonian* $\hat{H}$, *one finds*

$$\langle\Psi|\hat{H}|\Psi\rangle \geq E_{gs} , \tag{6.46}$$

where E_{gs} *is the ground-state energy of the system and the equality holds only if* $|\Psi\rangle = |\Psi_{gs}\rangle$ *with* $|\Psi_{gs}\rangle$ *ground-state of the system, i.e. such that* $\hat{H}|\Psi_{gs}\rangle = E_{gs}|\Psi_{gs}\rangle$.

Proof The many-body Hamiltonian $\hat{H}$ satisfies the exact eigenvalue problem

$$\hat{H}|\Psi_\alpha\rangle = E_\alpha|\Psi_\alpha\rangle , \tag{6.47}$$

where E_α are the ordered eigenvalues, i.e. such that $E_0 < E_1 < E_2 < ...$ with $E_0 = E_{gs}$, and $|\Psi_\alpha\rangle$ the corresponding orthonormalized eigenstates, i.e. such that $\langle\Psi_\alpha|\Psi_\beta\rangle = \delta_{\alpha,\beta}$ with $|\Psi_0\rangle = |\Psi_{gs}\rangle$. The generic many-body state $|\Psi\rangle$ can be written as

$$|\Psi\rangle = \sum_\alpha c_\alpha|\Psi_\alpha\rangle , \tag{6.48}$$

where c_α are the complex coefficients of the expansion such that

$$\sum_\alpha |c_\alpha|^2 = 1 . \tag{6.49}$$

Then one finds

$$\langle\Psi|\hat{H}|\Psi\rangle = \sum_{\alpha,\beta} c_\alpha^* c_\beta \langle\Psi_\alpha|\hat{H}|\Psi_\beta\rangle = \sum_{\alpha,\beta} c_\alpha^* c_\beta E_\beta \langle\Psi_\alpha|\Psi_\beta\rangle = \sum_{\alpha,\beta} c_\alpha^* c_\beta E_\beta \delta_{\alpha,\beta}$$

$$= \sum_\alpha |c_\alpha|^2 E_\alpha \geq \sum_\alpha |c_\alpha|^2 E_0 = E_0 = E_{gs} . \tag{6.50}$$

Obviously, the equality holds only if $c_0 = 1$ and, consequently, all the other coefficients are zero.

In 1927 by Douglas Hartree and Vladimir Fock used the variational principle to develop a powerful method for the study of interacting identical particles. We shall analyze this variational method in the following subsections.

6.3.2 Hartree for Bosons

In the case of N identical interacting bosons the Hartree approximation is simply given by

$$\Psi(x_1, x_2, ..., x_N) = \phi(x_1)\,\phi(x_2)\,...\,\phi(x_N)\,, \tag{6.51}$$

where the single-particle wavefunction $\phi(x)$ is unknown and it must be determined in a self-consistent way. Notice that, as previously stressed, this factorization implies that all particles belong to the same single-particle state, i.e. we are supposing that the interacting system is a pure Bose-Einstein condensate. This is a quite strong assumption, that is however reliable in the description of ultracold and dilute gases made of bosonic alkali-metal atoms (in 2001 Eric Cornell, Carl Weiman, and Wolfang Ketterle got the Nobel Prize in Physics for their experiments with these quantum gases), and which must be relaxed in the case of strongly-interacting bosonic systems (like superfluid ^{4}He). In the variational spirit of the Hartree approach the unknown wavefunction $\phi(x)$ is determined by minimizing the expectation value of the total Hamiltonian, given by

$$\langle \Psi | \hat{H} | \Psi \rangle = \int dx_1 dx_2 ... dx_N \, \Psi^*(x_1, x_2, ..., x_N) \hat{H} \Psi(x_1, x_2, ..., x_N)\,, \tag{6.52}$$

with respect to $\phi(x)$. In fact, by using Eq. (6.45) one finds immediately

$$\langle \Psi | \hat{H} | \Psi \rangle = N \int dx\, \phi^*(x) \hat{h}(x) \phi(x) + \frac{1}{2} N(N-1) \int dx\, dx'\, |\phi(x)|^2 V(x, x') |\phi(x')|^2\,, \tag{6.53}$$

which is a nonlinear energy functional of the single-particle wavefunction $\phi(x)$. It is called single-orbital Hartree functional for bosons. In this functional the first term is related to the single-particle hamiltonian $\hat{h}(x)$ while the second term is related to the inter-particle interaction potential $V(x, x')$. We minimize this functional with the following constraint due to the normalization

$$\int dx\, |\phi(x)|^2 = 1\,. \tag{6.54}$$

We get immediately the so-called Hartree equation for bosons

$$\left[\hat{h}(x) + U_{mf}(x)\right] \phi(x) = \epsilon\,\phi(x)\,, \tag{6.55}$$

where the mean-field potential $U_{mf}(x)$ reads

$$U_{mf}(x) = (N-1) \int dx'\, V(x, x')\, |\phi(x')|^2 \tag{6.56}$$

and ϵ is the Lagrange multiplier fixed by the normalization. It is important to observe that the mean-field potential $U_{mf}(x)$ depends on $\phi(x)$ and it must be obtained self-consistently. In other words, the Hartree equation of bosons is a integro-differential nonlinear Schrödinger equation whose nonlinear term gives the mean-field potential of the system.

In the case of spinless bosons, where $|x\rangle = |\mathbf{r}\rangle$, given the local bosonic density

$$\rho(\mathbf{r}) = N|\phi(\mathbf{r})|^2 , \tag{6.57}$$

under the assumption of a large number N of particles the Hartree variational energy reads

$$\langle \Psi|\hat{H}|\Psi\rangle = N \int d^3\mathbf{r} \, \phi^*(\mathbf{r}) \left[-\frac{\hbar^2}{2m}\nabla^2 + U(\mathbf{r}) \right] \phi(\mathbf{r}) + \frac{1}{2} \int d^3\mathbf{r} \, d^3\mathbf{r}' \, \rho(\mathbf{r})V(\mathbf{r} - \mathbf{r}')\rho(\mathbf{r}')$$

$$\tag{6.58}$$

while the Hartree equation becomes

$$\left[-\frac{\hbar^2}{2m}\nabla^2 + U(\mathbf{r}) + \int d^3\mathbf{r}' \, V(\mathbf{r} - \mathbf{r}') \, \rho(\mathbf{r}') \right] \phi(\mathbf{r}) = \epsilon \, \phi(\mathbf{r}) . \tag{6.59}$$

To conclude this subsection, we observe that in the case of a contact inter-particle potential, i.e.

$$V(\mathbf{r} - \mathbf{r}') = g \, \delta(\mathbf{r} - \mathbf{r}') , \tag{6.60}$$

the previous Hartree variational energy becomes

$$\langle \Psi|\hat{H}|\Psi\rangle = N \int d^3\mathbf{r} \, \phi^*(\mathbf{r}) \left[-\frac{\hbar^2}{2m}\nabla^2 + U(\mathbf{r}) \right] \phi(\mathbf{r}) + \frac{g}{2} \int d^3\mathbf{r} \, \rho(\mathbf{r})^2 \tag{6.61}$$

and the corresponding Hartee equation reads

$$\left[-\frac{\hbar^2}{2m}\nabla^2 + U(\mathbf{r}) + g \, \rho(\mathbf{r}) \right] \phi(\mathbf{r}) = \epsilon \, \phi(\mathbf{r}) , \tag{6.62}$$

which is the so-called Gross-Pitaevskii equation, deduced in 1961 by Eugene Gross and Lev Pitaevskii.

6.3.3 Hartree-Fock for Fermions

In the case of N identical interacting fermions, the approximation developed by Hartree and Fock is based on the Slater determinant we have seen previously, namely

$$\Psi(x_1, x_2, ..., x_N) = \frac{1}{\sqrt{N!}} \begin{pmatrix} \phi_1(x_1) & \phi_1(x_2) & ... & \phi_1(x_N) \\ \phi_2(x_1) & \phi_2(x_2) & ... & \phi_2(x_N) \\ ... & ... & ... & ... \\ \phi_N(x_1) & \phi_N(x_2) & ... & \phi_N(x_N) \end{pmatrix}, \tag{6.63}$$

where now the single-particle wavefunctions $\phi_n(x)$ are unknown and they are determined with a variational procedure. In fact, in the Hartree-Fock approach the unknown wavefunctions $\phi_n(x)$ are obtained by minimizing the expectation value of the total Hamiltonian, given by

$$\langle \Psi | \hat{H} | \Psi \rangle = \int dx_1 \, dx_2 \, ... \, dx_N \, \Psi^*(x_1, x_2, ..., x_N) \hat{H} \Psi(x_1, x_2, ..., x_N), \tag{6.64}$$

with respect to the N single-particle wavefunctions $\phi_n(x)$. By using Eq. (6.45) and after some tedious calculations one finds

$$\langle \Psi | \hat{H} | \Psi \rangle = \sum_{i=1}^{N} \int dx \, \phi_i^*(x) \hat{h}(x) \phi_i(x) + \frac{1}{2} \sum_{\substack{i,j=1 \\ i \neq j}}^{N} \left[\int dx \, dx' \, |\phi_i(x)|^2 V(x, x') |\phi_j(x')|^2 \right.$$
$$\left. - \int dx \, dx' \, \phi_i^*(x) \phi_j(x) V(x, x') \phi_j^*(x') \phi_i(x') \right], \tag{6.65}$$

which is a nonlinear energy functional of the N single-particle wavefunctions $\phi_i(x)$. In this functional the first term is related to the single-particle Hamiltonian $\hat{h}(x)$ while the second and the third terms are related to the inter-particle interaction potential $V(x, x')$. The second term is called direct term of interaction and the third term is called exchange term of interaction. We minimize this functional with the following constraints due to the normalization

$$\int dx \, |\phi_i(x)|^2 = 1, \qquad i = 1, 2, ..., N, \tag{6.66}$$

where, in the case of spin $1/2$ particles, one has

$$\phi_i(x) = \phi_i(\mathbf{r}, \sigma) = \tilde{\phi}_i(\mathbf{r}, \sigma) \chi_\sigma \tag{6.67}$$

with χ_σ the two-component spinor, and the integration over x means

$$\int dx = \int d^3\mathbf{r} \sum_{\sigma=\uparrow,\downarrow}, \tag{6.68}$$

such that

$$\int dx \, |\phi_i(x)|^2 = \int d^3\mathbf{r} \sum_{\sigma=\uparrow,\downarrow} |\phi_i(\mathbf{r}, \sigma)|^2 = \int d^3\mathbf{r} \sum_{\sigma=\uparrow,\downarrow} |\tilde{\phi}_i(\mathbf{r}, \sigma)|^2, \tag{6.69}$$

because $\chi_\sigma^* \chi_\sigma = 1$ and more generally $\chi_\sigma^* \chi_{\sigma'} = \delta_{\sigma,\sigma'}$.

After minimization of the energy functional we get the so-called Hartree-Fock equations

$$\left[\hat{h}(x) + \hat{U}_{mf}(x)\right]\phi_i(x) = \epsilon_i\,\phi_i(x) \tag{6.70}$$

where ϵ_i are the Lagrange multipliers fixed by the normalization and $\hat{U}_{mf}$ is a nonlocal mean-field operator. This nonlocal operator is given by

$$\hat{U}_{mf}(x)\,\phi_i(x) = U_d(x)\,\phi_i(x) - \sum_{j=1}^{N} U_x^{ji}(x)\phi_j(x)\,, \tag{6.71}$$

where the direct mean-field potential $U_d(x)$ reads

$$U_d(x) = \sum_{\substack{j=1 \\ j\neq i}}^{N} \int dx'\, V(x, x')\,|\phi_j(x')|^2\,, \tag{6.72}$$

while the exchange mean-field potential $U_x^{ji}(x)$ is instead

$$U_x^{ji}(x) = \int dx'\, \phi_j^*(x')\, V(x, x')\, \phi_i(x')\,. \tag{6.73}$$

If one neglects the exchange term, as done by Hartree in his original derivation, the so-called Hartree equations

$$\left[\hat{h}(x) + U_d(x)\right]\phi_i(x) = \epsilon_i\,\phi_i(x)\,, \tag{6.74}$$

are immediately derived. It is clearly much simpler to solve the Hartree equations instead of the Hartree-Fock ones. For this reason, in many applications the latter are often used. In the case of spin $1/2$ fermions, given the local fermionic density

$$\rho(\mathbf{r}) = \sum_{\sigma=\uparrow,\downarrow} \sum_{i=1}^{N} |\phi_i(\mathbf{r}, \sigma)|^2 = \sum_{\sigma=\uparrow,\downarrow} \rho(\mathbf{r}, \sigma)\,, \tag{6.75}$$

under the assumption of a large number N of particles the Hartree (direct) variational energy reads

$$E_D = \sum_{i=1}^{N} \sum_{\sigma=\uparrow,\downarrow} \int d^3\mathbf{r}\, \phi_i^*(\mathbf{r}, \sigma) \left[-\frac{\hbar^2}{2m}\nabla^2 + U(\mathbf{r})\right]\phi_i(\mathbf{r}, \sigma)$$

$$+ \frac{1}{2} \sum_{\sigma,\sigma'=\uparrow,\downarrow} \int d^3\mathbf{r}\, d^3\mathbf{r}'\, \rho(\mathbf{r}, \sigma)V(\mathbf{r} - \mathbf{r}')\rho(\mathbf{r}', \sigma') \tag{6.76}$$

and the corresponding Hartree equation becomes

$$\left[-\frac{\hbar^2}{2m}\nabla^2 + U(\mathbf{r}) + \sum_{\sigma'=\uparrow,\downarrow} \int d^3\mathbf{r}' \, V(\mathbf{r}-\mathbf{r}') \, \rho(\mathbf{r}',\sigma') \right] \phi_i(\mathbf{r},\sigma) = \epsilon_i \, \phi_i(\mathbf{r},\sigma) \, .$$

(6.77)

The Hartree-Fock variational energy is slightly more complex because it includes also the exchange energy, given by

$$E_X = -\frac{1}{2} \sum_{\substack{i,j=1 \\ i\neq j}}^{N} \sum_{\sigma=\uparrow,\downarrow} \int d^3\mathbf{r}\, d^3\mathbf{r}' \, \phi_i^*(\mathbf{r},\sigma)\phi_i(\mathbf{r}',\sigma)V(\mathbf{r}-\mathbf{r}')\phi_j^*(\mathbf{r}',\sigma)\phi_j(\mathbf{r},\sigma) \, .$$

(6.78)

Notice that in the exchange energy all the terms with opposite spins are zero due to the scalar product of spinors: $\chi_\sigma^* \chi_{\sigma'} = \delta_{\sigma,\sigma'}$. The existence of this exchange energy E_X is a direct consequence of the anti-symmetry of the many-body wave function, namely a consequence of the fermionic nature of the particles we are considering. Historically, this term E_X was obtained by Vladimir Fock to correct the first derivation of Douglas Hartree who used a not anti-symmetrized many-body wavefunction.

To conclude this subsection, we observe that in the case of a contact inter-particle potential, i.e.

$$V(\mathbf{r}-\mathbf{r}') = g\, \delta(\mathbf{r}-\mathbf{r}') \, ,$$

(6.79)

the Hartee-Fock (direct plus exchange) variational energy reads

$$E = \sum_{i=1}^{N} \sum_{\sigma=\uparrow,\downarrow} \int d^3\mathbf{r} \, \phi_i^*(\mathbf{r},\sigma) \left[-\frac{\hbar^2}{2m}\nabla^2 + U(\mathbf{r}) \right] \phi_i(\mathbf{r},\sigma)$$

(6.80)

$$+ \frac{g}{2} \sum_{\substack{i,j=1 \\ i\neq j}}^{N} \sum_{\sigma,\sigma'=\uparrow,\downarrow} \int d^3\mathbf{r} \left[|\phi_i(\mathbf{r},\sigma)|^2 |\phi_j(\mathbf{r},\sigma')|^2 - |\phi_i(\mathbf{r},\sigma)|^2 |\phi_j(\mathbf{r},\sigma)|^2 \delta_{\sigma,\sigma'} \right] \, .$$

It follows immediately that identical spin-polarized fermions with contact interaction are effectively non-interacting because in this case the interaction terms of direct and exchange energy exactly compensate to zero.

6.3.4 Mean-Field Approximation

A common strategy to solve the interacting many-body problem is the so-called mean-field approximation, which corresponds on finding a suitable mean-field potential which makes the system quasi-separable. The idea is the following. The interaction Hamiltonian

$$\hat{H}_I = \frac{1}{2} \sum_{\substack{i,j=1 \\ i \neq j}}^{N} V(x_i, x_j) \tag{6.81}$$

can be formally written as

$$\hat{H}_I = \hat{H}_{mf} + \left(\hat{H}_I - \hat{H}_{mf} \right) = \hat{H}_{mf} + H_{res} , \tag{6.82}$$

where $\hat{H}_{mf}$ is an appropriate separable mean-field Hamiltonian, i.e. such that

$$\hat{H}_{mf} = \sum_{i=1}^{N} U_{mf}(x_i) , \tag{6.83}$$

and $\hat{H}_{res}$ is the residual Hamiltonian. In this way the total Hamiltonian reads

$$\hat{H} = \hat{H}_0 + \hat{H}_{mf} + \hat{H}_{res} = \hat{H}' + \hat{H}_{res} , \tag{6.84}$$

where the effective Hamiltonian $\hat{H}'$ is given by

$$\hat{H}' = \hat{H}_0 + \hat{H}_{mf} = \sum_{i=1}^{N} \left[\hat{h}(x_i) + U_{mf}(x_i) \right] = \sum_{i=1}^{N} \left[-\frac{\hbar^2}{2m} \nabla_i^2 + U(x_i) + U_{mf}(x_i) \right] . \tag{6.85}$$

Neglecting the residual Hamiltonian $\hat{H}_{res}$, the system is described by the effective Hamiltonian $\hat{H}'$ and the N-body problem is reduced to the solution of the 1-body problem

$$\left[\hat{h}(x) + U_{mf}(x) \right] \phi_n(x) = \epsilon_n \, \phi_n(x) . \tag{6.86}$$

Obviously, the main difficulty of this approach is to determine the mean-field potential $U_{mf}(x)$ which minimizes the effect of the residual interaction $\hat{H}_{res}$. The mean-field potential $U_{mf}(x)$ can be chosen on the basis of the properties of the specific physical system or on the basis of known experimental data.

As shown in the previous subsections, $U_{mf}(x)$ can be also obtained by using the Hartree-Fock method. In the case of the Hartree-Fock method for bosons, introducing the effective Hamiltonian

$$\hat{H}' = \sum_{i=1}^{N} \left[\hat{h}(x_i) + U_{mf}(x_i) \right] \tag{6.87}$$

with $\hat{U}_{mf}(x_i)$ given by Eq. (6.56), it is immediate to verify that the bosonic many-body wavefunction $\Psi(x_1, x_2, ..., x_N)$ of Eq. (6.51), with the single-particle wavefunction $\phi(x)$ satisfying Eq. (6.55), is the lowest eigenfunction of $\hat{H}'$, and such that

$$\hat{H}'\Psi(x_1, x_2, ..., x_N) = N\epsilon\,\Psi(x_1, x_2, ..., x_N)\,. \tag{6.88}$$

On the other hand, $\Psi(x_1, x_2, ..., x_N)$ is not an eigenfunction of the total Hamiltonian $\hat{H}$, but it is an approximate variational wavefunction of the ground-state of the full Hamiltonian $\hat{H}$.

In a similar way, in the case of the Hartree-Fock method for fermions, introducing again the effective Hamiltonian

$$\hat{H}' = \sum_{i=1}^{N}\left[\hat{h}(x_i) + \hat{U}_{mf}(x_i)\right] \tag{6.89}$$

with $\hat{U}_{mf}(x_i)$ given by Eq. (6.71), it is not difficult to verify that the fermionic many-body wavefunction $\Psi(x_1, x_2, ..., x_N)$ of Eq. (6.63), with the single-particle wave-function $\phi(x)$ satisfying Eq. (6.70), is the lowest eigenfunction of $\hat{H}'$, and such that

$$\hat{H}'\Psi(x_1, x_2, ..., x_N) = \sum_{i=1}^{N}\epsilon_i\,\Psi(x_1, x_2, ..., x_N)\,. \tag{6.90}$$

On the other hand, $\Psi(x_1, x_2, ..., x_N)$ is not an eigenfunction of the total Hamiltonian $\hat{H}$, but it is an approximate variational wavefunction of the ground-state of the full Hamiltonian $\hat{H}$.

6.4 Density Functional Theory

In 1927 Llewellyn Thomas and Enrico Fermi independently proposed a statistical approach to the electronic structure of atoms with a large number N of electrons, which avoids tackling the solution of the many-body Schödinger equation by focusing on the electron local total number density $\rho(\mathbf{r})$. The starting point is the (kinetic) energy density of a uniform ideal gas of electrons at zero temperature in the thermodynamic limit, given by

$$\mathcal{E}_{ideal} = \frac{3}{5}\frac{\hbar^2}{2m}(3\pi^2)^{2/3}\rho^{5/3}\,, \tag{6.91}$$

where ρ is the uniform number density of electrons. Thomas and Fermi considered N electrons in a atom with nuclear charge Ze, such that

$$V(\mathbf{r} - \mathbf{r}') = \frac{e^2}{4\pi\varepsilon_0|\mathbf{r} - \mathbf{r}'|}\,. \tag{6.92}$$

the electron-electron Coulomb potential, while

$$U(\mathbf{r}) = -\frac{Ze^2}{4\pi\varepsilon_0|\mathbf{r}|} \tag{6.93}$$

is the nucleus-electron Coulomb potential with the nucleus centered at the orgin of the reference frame. Thomas and Fermi supposed that the ground-state energy of the electrons can be captured by the functional

$$E_{TF}[\rho] = T_{TF}[\rho] + E_D[\rho] + \int d^3\mathbf{r}\, U(\mathbf{r})\, \rho(\mathbf{r})\,, \tag{6.94}$$

where

$$T_{TF}[\rho] = \int d^3\mathbf{r}\, \frac{3}{5}\frac{\hbar^2}{2m}(3\pi^2)^{2/3}\rho(\mathbf{r})^{5/3} \tag{6.95}$$

is the Thomas-Fermi kinetic energy, i.e. the local approximation to the kinetic energy of the ideal Fermi gas,

$$E_D[\rho] = \frac{1}{2}\int d^3\mathbf{r}\, d^3\mathbf{r}'\, \rho(\mathbf{r})\, V(\mathbf{r}-\mathbf{r}')\, \rho(\mathbf{r}')\,, \tag{6.96}$$

is the (Hartree-like) direct energy of electron-electron interaction, and the third term in Eq. (6.94) is the energy of the nucleus-electron interaction.

It is important to stress that within this formalism the electron-nucleus potential $U(\mathbf{r})$ could be more general, taking into account the effect of many nucleons. The functional is minimized under the constraints $\rho(\mathbf{r}) \geq 0$ and

$$N = \int d^3\mathbf{r}\, \rho(\mathbf{r}) \tag{6.97}$$

to find the equilibrium condition

$$\frac{\hbar^2}{2m}(3\pi^2)^{2/3}\rho^{2/3} + U(\mathbf{r}) + U_{mf}(\mathbf{r}) = \mu\,, \tag{6.98}$$

where

$$U_{mf}(\mathbf{r}) = \int d^3\mathbf{r}'\, V(\mathbf{r}-\mathbf{r}')\, \rho(\mathbf{r}') \tag{6.99}$$

is the Hartree-like mean-field potential acting on the electrons and μ is the Lagrange multiplier fixed by the normalization. The previous equation can be written as

$$\rho(\mathbf{r}) = \frac{(2m)^{3/2}}{3\pi^2\hbar^3}\left(\mu - U(\mathbf{r}) - \int d^3\mathbf{r}'\, V(\mathbf{r}-\mathbf{r}')\, \rho(\mathbf{r}')\right)^{3/2} \tag{6.100}$$

Equation (6.100) is an implicit integral equation for the local electronic density $\rho(\mathbf{r})$ which can be solved numerically by using an iterative procedure.

Although this was an important first step, the Thomas-Fermi method is limited because the resulting kinetic energy is only approximate, and also because the method does not include the exchange energy of the electrons due to the Pauli principle. A better theoretical description is obtained with the Thomas-Fermi-Dirac-von Weizsäcker model, whose energy functional is given by

$$E_{TFDW}[\rho] = E_{TF}[\rho] + E_X[\rho] + E_W[\rho] , \tag{6.101}$$

where $E_{TF}[\rho]$ is the Thomas-Fermi energy functional of Eq. (6.94),

$$E_X[\rho] = -\int d^3\mathbf{r} \, \frac{3}{4} \frac{e^2}{4\pi\varepsilon_0} (\frac{3}{\pi})^{1/3} \rho(\mathbf{r})^{4/3} \tag{6.102}$$

is the functional correction introduced by Paul Dirac in 1930 to take into account the exchange energy which appears in the Hartree-Fock method, and

$$E_W[\rho] = \int d^3\mathbf{r} \, \lambda \frac{\hbar^2}{2m} (\nabla\sqrt{\rho(\mathbf{r})})^2 \tag{6.103}$$

is the functional correction introduced by Carl Friedrich von Weizsäcker in 1953 to improve the accuracy of the kinetic energy with λ an adjustable parameter. Notice that a simple dimensional analysis shows that the Thomas-Fermi kinetic energy density must scale as $\hbar^2 \rho^{5/3}/(2m)$, while the exchange energy density must scale as $e^2 \rho^{4/3}/(4\pi\varepsilon_0)$.

In 1964 the density functional approach was put on a firm theoretical footing by Pierre Hohenberg and Walter Kohn. We follow their original reasoning which does not take into account explicitly the spin degrees. First of all they observed that, given a many-body wavefunction $\Psi(\mathbf{r}_1, \mathbf{r}_2, ..., \mathbf{r}_N)$, the associated one-body density $\rho(\mathbf{r})$ reads

$$\rho(\mathbf{r}) = \langle\Psi| \sum_{i=1}^{N} \delta(\mathbf{r} - \mathbf{r}_i)|\Psi\rangle = N \int d^3\mathbf{r}_2 ... d^3\mathbf{r}_N \, |\Psi(\mathbf{r}, \mathbf{r}_2, ..., \mathbf{r}_N)|^2 . \tag{6.104}$$

Then they formulated two rigorous theorems for a system of identical particles (bosons or fermions) described by the Hamiltonian

$$\hat{H} = \hat{T} + \hat{U} + \hat{V} \tag{6.105}$$

where

$$\hat{T} = \sum_{i=1}^{N} -\frac{\hbar^2}{2m} \nabla_i^2 \tag{6.106}$$

is the many-body kinetic energy operator,

$$\hat{U} = \sum_{i=1}^{N} U(\mathbf{r}_i) \tag{6.107}$$

is the many-body external potential operator, and

$$\hat{V} = \frac{1}{2} \sum_{\substack{i,j=1 \\ i \neq j}}^{N} V(\mathbf{r}_i, \mathbf{r}_j) \tag{6.108}$$

is the many-body interaction potential operator. We now state the two theorems of Hohenberg and Kohn, which are based on the variational principle discussed in the previous section, but first of all we observe that for any $\Psi(\mathbf{r}_1, \mathbf{r}_2, ..., \mathbf{r}_N)$ one has

$$\langle \Psi | \hat{H} | \Psi \rangle = \langle \Psi | \hat{T} + \hat{V} | \Psi \rangle + \langle \Psi | \hat{U} | \Psi \rangle = \langle \Psi | \hat{T} + \hat{V} | \Psi \rangle + \int U(\mathbf{r}) \, \rho(\mathbf{r}) \, d^3\mathbf{r} \, . \tag{6.109}$$

This formula shows explicitly that the external energy $\langle \Psi | \hat{U} | \Psi \rangle$ is a functional of the local density $\rho(\mathbf{r})$. The Hohemberg-Kohn theorems ensure that also the internal energy $\langle \Psi | \hat{F} | \Psi \rangle = \langle \Psi | \hat{T} + \hat{V} | \Psi \rangle$ is a functional of $\rho(\mathbf{r})$.

Theorem 1 *For a system of N identical interacting particles in an external potential $U(\mathbf{r})$ the density $\rho_{gs}(\mathbf{r})$ of the non degenerate ground-state $\Psi_{gs}(\mathbf{r}_1, \mathbf{r}_2, ..., \mathbf{r}_N)$ is uniquely determined by the external potential. In other words, there is a one-to-one correspondence between $U(\mathbf{r})$ and $\rho_{gs}(\mathbf{r})$.*

Proof The proof of this theorem proceeds by reductio ad absurdum. Let there be two different external potentials, $U_1(\mathbf{r})$ and $U_2(\mathbf{r})$, that give rise to the same ground-state density $\rho_{gs}(\mathbf{r})$. Due to the variational principle, the associated Hamiltonians, $\hat{H}_1 = \hat{F} + \hat{U}_1$ and $\hat{H}_2 = \hat{F} + \hat{U}_2$ have different ground-state wavefunctions, $\Psi_{gs,1}(\mathbf{r}_1, \mathbf{r}_2, ..., \mathbf{r}_N)$ and $\Psi_{gs,2}(\mathbf{r}_1, \mathbf{r}_2, ..., \mathbf{r}_N)$, that each yield $\rho_{gs}(\mathbf{r})$. Using the variational principle

$$\begin{aligned} E_{gs,1} &< \langle \Psi_{gs,2} | \hat{H}_1 | \Psi_{gs,2} \rangle = \langle \Psi_{gs,2} | \hat{H}_2 | \Psi_{gs,2} \rangle + \langle \Psi_{gs,2} | \hat{H}_1 - \hat{H}_2 | \Psi_{gs,2} \rangle \\ &= E_{gs,2} + \int \rho_{gs}(\mathbf{r}) [U_1(\mathbf{r}) - U_2(\mathbf{r})] \, d^3\mathbf{r} \, , \end{aligned} \tag{6.110}$$

where $E_{gs,1}$ and $E_{gs,2}$ are the ground-state energies of $\hat{H}_1$ and $\hat{H}_2$ respectively. An equivalent expression for Eq. (6.110) holds when the subscripts are interchanged, namely

$$\begin{aligned} E_{gs,2} &< \langle \Psi_{gs,1} | \hat{H}_2 | \Psi_{gs,1} \rangle = \langle \Psi_{gs,1} | \hat{H}_1 | \Psi_{gs,1} \rangle + \langle \Psi_{gs,1} | \hat{H}_2 - \hat{H}_1 | \Psi_{gs,1} \rangle \\ &= E_{gs,1} + \int \rho_{gs}(\mathbf{r}) [U_2(\mathbf{r}) - U_1(\mathbf{r})] \, d^3\mathbf{r} \, . \end{aligned} \tag{6.111}$$

Therefore adding the inequality (6.110) to the inequality (6.110) leads to the result:

$$E_{gs,1} + E_{gs,2} < E_{gs,2} + E_{gs,1} \,, \tag{6.112}$$

which is a contradiction.

From this theorems one deduces immediately two corollaries.

Corollary 1 *The many-body ground-state wavefunction* $\Psi_{gs}(\mathbf{r}_1, \mathbf{r}_2, ..., \mathbf{r}_N)$ *is a bijective function of* $\rho_{gs}(\mathbf{r})$.

Proof As just shown in Theorem 1, $\rho_{gs}(\mathbf{r})$ determines $U(\mathbf{r})$, and $U(\mathbf{r})$ determines $\hat{H}$ and therefore $\Psi_{gs}(\mathbf{r}_1, \mathbf{r}_2, ..., \mathbf{r}_N)$. This means that $\Psi_{gs}(\mathbf{r}_1, \mathbf{r}_2, ..., \mathbf{r}_N)$ is a bijective function of $\rho_{gs}(\mathbf{r})$. Symbolically we can write

$$\Psi_{gs} = G[\rho_{gs}] \,. \tag{6.113}$$

Corollary 2 *The internal energy* $\langle\Psi_{gs}|\hat{F}|\Psi_{gs}\rangle$ *of the ground state is a universal functional* $F[\rho]$ *of* $\rho_{gs}(\mathbf{r})$.

Proof E_{gs} is a functional of $\Psi_{gs}(\mathbf{r}_1, \mathbf{r}_2, ..., \mathbf{r}_N)$. Symbolically we have

$$E_{gs} = \Phi[\Psi_{gs}] \,. \tag{6.114}$$

But we have just seen that $\Psi_{gs}(\mathbf{r}_1, \mathbf{r}_2, ..., \mathbf{r}_N)$ is a bijective function of $\rho_{gs}(\mathbf{r})$, in particular

$$\Psi_{gs} = G[\rho_{gs}] \,. \tag{6.115}$$

Consequently E_{gs} is a functional of $\rho_{gs}(\mathbf{r})$, namely

$$E_{gs} = \Phi[G[\rho_{gs}]] = E[\rho_{gs}] \,. \tag{6.116}$$

Due to the simple structure of the external energy, the functional $E[\rho]$ is clearly given by

$$E[\rho] = F[\rho] + \int d^3\mathbf{r}\, U(\mathbf{r})\, \rho(\mathbf{r}) \,, \tag{6.117}$$

where $F[\rho]$ is a universal functional which describes the internal (kinetic and interaction) energy.

Theorem 2 *For a system of N identical interacting particles in an external potential $U(\mathbf{r})$ the density functional*

$$E[\rho] = F[\rho] + \int d^3\mathbf{r}\, U(\mathbf{r})\, \rho(\mathbf{r})$$

is such that $E[\rho] \geq E[\rho_{gs}] = E_{gs}$ for any trial density $\rho(\mathbf{r})$, and the equality holds only for $\rho(\mathbf{r}) = \rho_{gs}(\mathbf{r})$.

Proof We have seen that E_{gs} is a functional of $\Psi_{gs}(\mathbf{r}_1, \mathbf{r}_2, ..., \mathbf{r}_N)$ but also a functional of $\rho_{gs}(\mathbf{r})$, i.e.

$$E_{gs} = \Phi[\Psi_{gs}] = \Phi[G[\rho_{gs}]] = E[\rho_{gs}] . \tag{6.118}$$

If $\Psi(\mathbf{r}_1, \mathbf{r}_2, ..., \mathbf{r}_N) \neq \Psi_{gs}(\mathbf{r}_1, \mathbf{r}_2, ..., \mathbf{r}_N)$ from the variational principle we have

$$\Phi[\Psi] = E[\rho] > E_{gs} , \tag{6.119}$$

where $\rho(\mathbf{r})$ is a local density different from $\rho_{gs}(\mathbf{r})$, due to Theorem 1. It follows immediately

$$E[\rho] > E[\rho_{gs}] , \tag{6.120}$$

if $\rho(\mathbf{r}) \neq \rho_{gs}(\mathbf{r})$.

The functional $F[\rho]$ is universal in the sense that it does not depend on the external potential $U(\mathbf{r})$ but only on the inter-particle potential $V(\mathbf{r} - \mathbf{r}')$, which is the familiar Coulomb potential in the case of electrons. Indeed, the description of electrons in atoms, molecules and solids is based on the choice of $U(\mathbf{r})$ while $F[\rho]$ is the same.

In the last forty years several approaches have been developed to find approximate but reliable expressions for $F[\rho]$ in the case of electrons. For instance, within the Thomas-Fermi-Dirac-von Weizsäcker model we have seen that

$$F[\rho] = T_{TF}[\rho] + E_D[\rho] + E_X[\rho] + E_W[\rho] . \tag{6.121}$$

Instead, for a Bose-Einstein condensate made of dilute and ultracold atoms, the bosonic Hatree-Fock method suggests

$$F[\rho] = E_W[\rho] + \frac{g}{2} \int d^3\mathbf{r}\, \rho(\mathbf{r})^2 \tag{6.122}$$

because $V(\mathbf{r} - \mathbf{r}') \simeq g\, \delta(\mathbf{r} - \mathbf{r}')$, where $g = 4\pi\hbar^2 a_s/m$, with m the atomic mass and a_s the inter-atomic s-wave scattering length. Note that on the basis of empirical observations in the $E_W[\rho]$ term one usually takes $\lambda = 1/6$ for electrons and $\lambda = 1$ for bosons.

Nowadays the most used density functional for electrons is the one proposed by Walter Kohn and Lu Jeu Sham in 1965. In the Kohn-Sham density functional approach the universal functional $F[\rho]$ is given by

$$F[\rho] = T_{KS}[\rho] + E_D[\rho] + E_{XC}[\rho] \tag{6.123}$$

where

$$T_{KS}[\rho] = \sum_{i=1}^{N} \phi_i^*(\mathbf{r}) \left(-\frac{\hbar^2}{2m} \nabla^2 \right) \phi_i(\mathbf{r}) \tag{6.124}$$

is the Kohn-Sham kinetic energy, where the orbitals $\phi_i(\mathbf{r})$ determine the local density, namely

$$\rho(\mathbf{r}) = \sum_{i=1}^{N} |\phi_i(\mathbf{r})|^2 , \tag{6.125}$$

the direct (or Hartree-like) energy of interaction $E_D[\rho]$ has the familiar form of Eq. (6.96), and $E_{XC}[\rho]$ is the so-called exchange-correlation energy, which simply takes into account the missing energy with respect to the exact result. The minimization of the Kohn-Sham density functional gives

$$\left[-\frac{\hbar^2}{2m} \nabla^2 + U(\mathbf{r}) + \int d^3\mathbf{r}' \, V(\mathbf{r} - \mathbf{r}') \, \rho(\mathbf{r}') + \frac{\delta E_{XC}[\rho]}{\delta \rho(\mathbf{r})} \right] \phi_i(\mathbf{r}) = \epsilon_i \, \phi_i(\mathbf{r}) \tag{6.126}$$

which are the local Kohn-Sham equations for the orbitals $\phi_i(\mathbf{r})$, with ϵ_i the Lagrange multipliers fixed by the normalization to one of the orbitals. Notice that the third term on the left side of the previous equation is obtained by using

$$\frac{\delta E_{XC}[\rho]}{\delta \phi_i^*(\mathbf{r})} = \frac{\delta E_{XC}[\rho]}{\delta \rho(\mathbf{r})} \frac{\delta \rho(\mathbf{r})}{\delta \phi_i^*(\mathbf{r})} = \frac{\delta E_{XC}[\rho]}{\delta \rho(\mathbf{r})} \phi_i(\mathbf{r}) . \tag{6.127}$$

In many applications the (usually unknown) exchange-correlation energy is written as

$$E_{XC}[\rho] = E_X[\rho] + E_C[\rho] , \tag{6.128}$$

where the correlation energy $E_C[\rho]$ is fitted from Monte Carlo calculations. Contrary to the Hartree-Fock theory, in the Kohn-Sham approach the single-particle orbitals $\phi_i(\mathbf{r})$ are not related to a many-body wavefunction but to the local density $\rho(\mathbf{r})$: this implies that the orbitals $\phi_i(\mathbf{r})$ and energies ϵ_i of the Kohn-Sham density functional have no deep physical meaning. Nevertheless, from the numerical point of view, the solution of the Kohn-Sham equations is much simpler than that of the Hartree-Fock ones, since, contrary to the effective Hartree-Fock mean-field potential, the effective Kohn-Sham mean-field potential is local.

It is important to stress that the density functional theory is an extremely useful approach for the description of atoms, molecules, and metals. Nowadays the success of density functional theory not only encompasses standard bulk materials but also complex materials such as proteins, carbon nanotubes, and nuclear matter.

6.5 Molecules and the Born-Oppenheimer Approximation

Up to now we have considered only single atoms with many-electrons. In this section we discuss systems composed by many atoms, and in particular molecules.

It is well known that a generic molecule is made of N_n atomic nuclei with electric charges $Z_\alpha e$ and masses M_α ($\alpha = 1, 2, ..., N_n$) and N_e electrons with charges $-e$ and masses m. Neglecting the finite structure of atomic nuclei, the Hamiltonian of the molecule can be written as

$$\hat{H} = \hat{H}_n + \hat{H}_e + V_{ne} , \tag{6.129}$$

where

$$\hat{H}_n = \sum_{\alpha=1}^{N_n} -\frac{\hbar^2}{2M_\alpha} \nabla_\alpha^2 + \frac{1}{2} \sum_{\substack{\alpha,\beta=1 \\ \alpha \neq \beta}}^{N_n} \frac{Z_\alpha Z_\beta e^2}{4\pi\varepsilon_0} \frac{1}{|\mathbf{R}_\alpha - \mathbf{R}_\beta|} \tag{6.130}$$

is the Hamiltonian of the atomic nuclei, with $\mathbf{R}_\alpha$ the position of the α-th nucleus,

$$\hat{H}_e = \sum_{i=1}^{N_e} -\frac{\hbar^2}{2m} \nabla_i^2 + \frac{1}{2} \sum_{\substack{i,j=1 \\ i \neq j}}^{N_e} \frac{e^2}{4\pi\varepsilon_0} \frac{1}{|\mathbf{r}_i - \mathbf{r}_j|} \tag{6.131}$$

is the Hamiltonian of the electrons, with $\mathbf{r}_i$ the position of the i-th electron, and

$$V_{ne} = -\sum_{\alpha,i=1}^{N_n,N_e} \frac{Z_\alpha e^2}{4\pi\varepsilon_0} \frac{1}{|\mathbf{R}_\alpha - \mathbf{r}_i|} \tag{6.132}$$

is the potential energy of the Coulomb interaction between atomic nuclei and electrons.

It is clear that the computation of the ground-state energy and the many-body wavefunction of an average-size molecule is a formidable task. For instance, the benzene molecule (C_6H_6) consists of 12 atomic nuclei and 42 electrons, and this means that its many-body wavefunction has $(12 + 42) \times 3 = 162$ variables: the spatial coordinates of the electrons and the nuclei. The exact many-body Schrodinger equation for the ground-state is given by

$$\hat{H}\Psi(R,r) = E\ \Psi(R,r) , \tag{6.133}$$

where $\Psi(R, r) = \Psi(\mathbf{R}_1, ..., \mathbf{R}_{N_n}, \mathbf{r}_1, ..., \mathbf{r}_{N_e})$ is the ground-state wavefunction, with $R = (\mathbf{R}_1, ..., \mathbf{R}_{N_n})$ and $r = (\mathbf{r}_1, ..., \mathbf{r}_{N_e})$ multi-vectors for nuclear and electronic coordinates respectively.

In 1927 Max Born and Julius Robert Oppenheimer suggested a reliable approximation to treat this problem. Their approach is based on the separation of the fast electron dynamics from the slow motion of the nuclei. In the so-called Born-Oppenheimer approximation the many-body wave function of the molecule is factorized as

$$\Psi(R, r) = \Psi_e(r; R) \, \Psi_n(R) , \qquad (6.134)$$

where $\Psi_n(R)$ is the nuclear wavefunction and $\Psi_e(r; R)$ is the electronic wavefunction, which depends also on nuclear coordinates. Moreover, one assumes that the electronic wavefunction $\Psi_e(r; R)$ is a solution of the following eigenvalue equation

$$\left(\hat{H}_e + V_{ne}\right) \Psi_e(r; R) = E_e(R)\Psi_e(r; R) . \qquad (6.135)$$

where the electronic eigenenergy $E_e(R)$ depends on the nuclear coordinates because of the nucleon-electron potential $V_{ne} = V_{ne}(R, r)$. This equation describes an electronic eigenstate compatible with a given geometrical configuration R of the atomic nuclei. Physically, it corresponds on assuming that nuclear and electronic motions are somehow decoupled. In this way one gets

$$
\begin{aligned}
\hat{H}\Psi_e(r; R) \, \Psi_n(R) &= \left(\hat{H}_n + \hat{H}_e + V_{ne}\right) \Psi_e(r; R) \, \Psi_n(R) \\
&= \hat{H}_n \Psi_e(r; R) \, \Psi_n(R) + \left(\hat{H}_e + V_{ne}\right) \Psi_e(r; R) \, \Psi_n(R) \\
&= \hat{H}_n \Psi_e(r; R) \, \Psi_n(R) + E_e(R)\Psi_e(r; R) \, \Psi_n(R) \\
&\simeq \Psi_e(r; R) \left(\hat{H}_n \Psi_n(R) + E_e(R)\right) \Psi_n(R) \qquad (6.136)
\end{aligned}
$$

where the last approximate equality $\simeq$ is sound because the effect of the kinetic term of the operator $\hat{H}_n$ on the electronic wavefunction $\Psi_e(r; R)$ is small, being the nuclear masses M_α much larger than the electronic mass m. Finally, from the previous equation and Eqs. (6.133) and (6.135) one finds

$$\left(\hat{H}_n + E_e(R)\right) \Psi_n(R) = E \, \Psi_n(R) , \qquad (6.137)$$

that is the equation of the adiabatic motion of atomic nuclei in the mean-field potential $E_e(R)$ generated by the electrons.

Equations (6.135) and (6.137) are still quite complex, and they are usually solved within some approximate quantum many-body method, e.g. the Hartree-Fock method or the density functional theory.

To conclude, we observe that in this section we have considered molecules, but the Born-Oppenheimer approximation is crucial in any context where there is more than a single atom around, which includes atomic gases, clusters, crystals, and many other physical systems.

6.6 Solved Problems

Problem 6.1

By using the Gaussian variational method calculate the approximate energy of the ground-state of a one-dimensional quantum particle under the action of the quartic potential

$$U(x) = A\,x^4\,.$$

Solution

The stationary Schrödinger equation of the particle is given by

$$\left[-\frac{\hbar^2}{2m}\frac{\partial^2}{\partial x^2} + A\,x^4\right]\psi(x) = \epsilon\,\psi(x)\,.$$

This equation can be seen as the Euler-Lagrange equation obtained by minimizing the energy functional

$$E = \int dx\,\psi^*(x)\left[-\frac{\hbar^2}{2m}\frac{\partial^2}{\partial x^2} + A\,x^4\right]\psi(x) = \int dx\left[\frac{\hbar^2}{2m}\left|\frac{\partial\psi(x)}{\partial x}\right|^2 + A\,x^4|\psi(x)|^2\right]\,,$$

with the normalization condition

$$\int dx\,|\psi(x)|^2 = 1\,.$$

According to the Gaussian variational method the wavefunction is supposed to be given by

$$\psi(x) = \frac{e^{-x^2/(2\sigma^2)}}{\pi^{1/4}\sigma^{1/2}}\,,$$

where σ is the variational parameter. Inserting this variational wavefunction in the energy functional and integrating over x we get

$$E = \frac{\hbar^2}{4m\sigma^2} + \frac{3}{4}A\,\sigma^4\,.$$

Minimizing this energy with respect to the variational parameter σ, i.e. setting $\frac{dE}{d\sigma} = 0$, we find

$$\sigma = \left(\frac{\hbar^2}{6mA}\right)^{1/6} .$$

Substituting this value of σ in the energy E we finally obtain

$$E = \frac{9}{4}\left(\frac{\hbar^2}{6m}\right)^{2/3} A^{1/3} ,$$

which is the approximate energy of the ground-state. This energy is surely larger or equal to the energy associated to the exact ground-state of the system.

Problem 6.2
Derive the Gross-Pitaevskii equation for a system of N Bose-condensed particles with contact interaction

$$V(\mathbf{r} - \mathbf{r}') = g\,\delta(\mathbf{r} - \mathbf{r}') ,$$

and prove that the Lagrange multiplier ϵ of the Gross-Pitaevskii equation is the chemical potential μ of the system.

Solution
The Hartree equation for N Bose-condensed particles is given by

$$\left[-\frac{\hbar^2}{2m}\nabla^2 + U(\mathbf{r}) + (N-1)\int d^3\mathbf{r}'\,|\psi(\mathbf{r}')|^2 V(\mathbf{r} - \mathbf{r}')\right]\psi(\mathbf{r}) = \epsilon\,\psi(\mathbf{r}) ,$$

where $\psi(\mathbf{r})$ is the wavefunction normalized to one. Inserting the contact potential we obtain

$$\left[-\frac{\hbar^2}{2m}\nabla^2 + U(\mathbf{r}) + (N-1)g|\psi(\mathbf{r})|^2\right]\psi(\mathbf{r}) = \epsilon\,\psi(\mathbf{r}) .$$

Under the reasonable hypotesis that $N \gg 1$ we get

$$\left[-\frac{\hbar^2}{2m}\nabla^2 + U(\mathbf{r}) + Ng|\psi(\mathbf{r})|^2\right]\psi(\mathbf{r}) = \epsilon\,\psi(\mathbf{r}) .$$

that is the Gross-Pitaevskii equation. This equation describes accurately a dilute Bose-Einstein condensate where the true inter-particle potential can be approximated with the contact potential. In this case the parameter g can be written as

$$g = \frac{4\pi\hbar^2 a_s}{m} ,$$

where a_s is the s-wave scattering length of the true inter-particle potential $V(\mathbf{r} - \mathbf{r})$.

The energy functional associated to the Gross-Pitaevskii equation is given by

$$E = N \int d^3\mathbf{r} \left\{ \psi^*(\mathbf{r}) \left[-\frac{\hbar^2}{2m}\nabla^2 + U(\mathbf{r}) \right] \psi(\mathbf{r}) + \frac{1}{2}Ng|\psi(\mathbf{r})|^4 \right\} \ .$$

The chemical potential is defined as

$$\mu = \frac{\partial E}{\partial N} \ .$$

On the basis of this definition we obtain

$$\mu = \int d^3\mathbf{r} \left\{ \psi^*(\mathbf{r}) \left[-\frac{\hbar^2}{2m}\nabla^2 + U(\mathbf{r}) \right] \psi(\mathbf{r}) + Ng|\psi(\mathbf{r})|^4 \right\} \ .$$

On the other hand, if we insert $\psi^*(\mathbf{r})$ on the left side of the Gross-Pitaevskii equation and we integrate over $\mathbf{r}$ we get

$$\int d^3\mathbf{r} \left\{ \psi^*(\mathbf{r}) \left[-\frac{\hbar^2}{2m}\nabla^2 + U(\mathbf{r}) \right] \psi(\mathbf{r}) + Ng|\psi(\mathbf{r})|^4 \right\} = \epsilon \ ,$$

because

$$\int d^3\mathbf{r}|\psi(\mathbf{r})|^2 = 1 \ .$$

By comparing the formulas of ϵ and μ we conclude that $\epsilon = \mu$. This result is also known as Koopmans theorem. Indeed, more rigorously the chemical potential defined as

$$\mu = E_N - E_{N-1} \ ,$$

but also in this case one easily finds that $\epsilon = \mu$ using the correct expression of the energy

$$E_N = N \int d^3\mathbf{r} \left\{ \psi^*(\mathbf{r}) \left[-\frac{\hbar^2}{2m}\nabla^2 + U(\mathbf{r}) \right] \psi(\mathbf{r}) + \frac{1}{2}(N-1)g|\psi(\mathbf{r})|^4 \right\} \ ,$$

where it appears $(N-1)$ instead of N in front of the interaction strength g.

Problem 6.3

Write the energy functional associated to the Gross-Pitaevskii equation as a functional of the local density $\rho(\mathbf{r})$ of the Bose-Einstein condensate.

Solution

We have seen that the energy functional associated to the Gross-Pitaevskii equation of the Bose-Einstein condensate is given by

$$E = N \int d^3\mathbf{r} \left\{ \psi^*(\mathbf{r}) \left[-\frac{\hbar^2}{2m} \nabla^2 + U(\mathbf{r}) \right] \psi(\mathbf{r}) + \frac{1}{2} Ng|\psi(\mathbf{r})|^4 \right\} ,$$

but it can also be written as

$$E = N \int d^3\mathbf{r} \left\{ \frac{\hbar^2}{2m} |\nabla\psi(\mathbf{r})|^2 + U(\mathbf{r})|\psi(\mathbf{r})|^2 + \frac{1}{2} Ng|\psi(\mathbf{r})|^4 \right\} ,$$

if the wavefunction is zero at the surface of integration volume. We now introduce the local density of the Bose-Einstein condensate as

$$\rho(\mathbf{r}) = N\psi(\mathbf{r})^2 ,$$

supposing that the wavefunction is real. In this way the energy functional can be immediately written as

$$E = \int d^3\mathbf{r} \left\{ \frac{\hbar^2}{2m} \left(\nabla\sqrt{\rho(\mathbf{r})} \right)^2 + U(\mathbf{r})\rho(\mathbf{r}) + \frac{1}{2} g\rho(\mathbf{r})^2 \right\} ,$$

that is a fuctional of the local density $\rho(\mathbf{r})$, as required.

Problem 6.4

By using the Gaussian variational method on the Gross-Pitaevskii functional calculate the energy per particle of a Bose-Einstein condensate under harmonic confinement, given by

$$U(\mathbf{r}) = \frac{1}{2} m\omega^2(x^2 + y^2 + z^2) .$$

Solution

We start from the Gaussian variational wave function

$$\psi(\mathbf{r}) = \frac{1}{\pi^{3/4} a_H^{3/2} \sigma^{3/2}} e^{-(x^2+y^2+z^2)/(2a_H^2\sigma^2)} ,$$

where $a_H = \sqrt{\hbar/(m\omega)}$ is the characteristic length of the harmonic confinement and σ is the dimensionless variational parameter. Inserting this wave function in the Gross-Pitaevskii energy functional

$$E = N \int d^3\mathbf{r} \left\{ \psi^*(\mathbf{r}) \left[-\frac{\hbar^2}{2m} \nabla^2 + \frac{1}{2} m\omega^2(x^2 + y^2 + z^2) \right] \psi(\mathbf{r}) + \frac{1}{2} Ng|\psi(\mathbf{r})|^4 \right\} ,$$

after integration we obtain the energy E as a function of σ, namely

$$E = N\hbar\omega \left(\frac{3}{4}\frac{1}{\sigma^2} + \frac{3}{4}\sigma^2 + \frac{\gamma}{2}\frac{1}{\sigma^3} \right) ,$$

where $\gamma = \sqrt{2/\pi}Na_s/a_H$ is the dimensionless strength of the inter-particle interaction with $g = 4\pi\hbar^2 a_s/m$ and a_s the s-wave scattering length. The best choice for σ is obtained by extremizing the energy function:

$$\frac{dE}{d\sigma} = 0 ,$$

from which we obtain

$$\sigma(\sigma^4 - 1) = \gamma .$$

It is clear that σ grows with γ, if $\gamma > 0$. Instead, if $\gamma < 0$ there are two possible values of σ: one value corresponds to a minimum of the energy E (meta-stable solution) and the other corresponds to a maximum of the energy E (unstable solution). It is straightforward to show that these two solutions exists only if $\gamma > -4/5^{5/4}$.

Further Reading

For the many-electron atom and the Hartree-Fock method:
B.H. Bransden and C.J. Joachain, Physics of Atoms and Molecules, Chap. 8, Sects. 8.1, 8.2 and 8.4 (Prentice Hall, Upper Saddle River, 2003).
For the density functional theory:
B.H. Bransden and C.J. Joachain, Physics of Atoms and Molecules, Chap. 8, Sects. 8.3 and 8.6 (Prentice Hall, Upper Saddle River, 2003).
E. Lipparini, Modern Many-Particle Physics: Atomic Gases, Quantum Dots and Quantum Fluids, Chap. 4, Sects. 4.1, 4.2, and 4.3 (World Scientific, Singapore, 2003).

Chapter 7
Second Quantization of Matter

In this chapter we discuss the second quantization of the non-relativistic matter field, that is the Schrödinger field. We show that the Schrödinger field can be expressed as a infinite sum of harmonic oscillators. These oscillators, which describe the possible eigenfrequencies of the matter field, are quantized by introducing creation and annihilation operators acting on the Fock space of number representation. We show how, depending on the commutation rule, these ladder operators can model accurately bosonic and fermionic particles which interact among themselves or with the quantum electromagnetic field. The second quantization (also called quantum field theory) is the powerful tool to describe phenomena, both at zero and finite temperature, where the number of particles is not conserved or it is conserved only on the average.

7.1 Schrödinger Field

In the second chapter we have seen that the light field is composed of an infinite number of quanta, the photons, as explained in 1927 by Paul Dirac. In the same year Eugene Wigner and Pascual Jordan proposed something similar for the matter. They suggested that in non-relativistic quantum mechanics the matter field is nothing but the single-particle Schrödinger field $\psi(\mathbf{r}, t)$ of quantum mechanics, which satisfies the time-dependent Schrödinger equation

$$i\hbar \frac{\partial}{\partial t} \psi(\mathbf{r}, t) = \left[-\frac{\hbar^2}{2m} \nabla^2 + U(\mathbf{r}) \right] \psi(\mathbf{r}, t) , \tag{7.1}$$

where $U(\mathbf{r})$ is the external potential acting on the quantum particle. Setting

$$\psi(\mathbf{r}, t) = \phi_\alpha(\mathbf{r}) \, e^{-i\epsilon_\alpha t/\hbar} \tag{7.2}$$

© Springer International Publishing AG 2017
L. Salasnich, *Quantum Physics of Light and Matter*, UNITEXT for Physics,
DOI 10.1007/978-3-319-52998-1_7

one finds

$$\left[-\frac{\hbar^2}{2m}\nabla^2 + U(\mathbf{r})\right]\phi_\alpha(\mathbf{r}) = \epsilon_\alpha \phi_\alpha(\mathbf{r}) \tag{7.3}$$

that is the stationary Schrödinger equation for the eigenfunction $\phi_\alpha(\mathbf{r})$ with eigenvalue ϵ_α where α is the label which represents a set of quantum numbers. In this case the Schrödinger field $\psi(\mathbf{r}, t)$ is a stationary state of the system, in fact

$$|\psi(\mathbf{r}, t)|^2 = |\phi_\alpha(\mathbf{r})|^2 \tag{7.4}$$

does not depend on time t.

In general, the Schrödinger field $\psi(\mathbf{r}, t)$ is not a stationary state of the system because the space dependence and time dependence cannot expressed as in Eq. (7.2), but, as suggested by Wigner and Jordan, one can surely expand the field as

$$\psi(\mathbf{r}, t) = \sum_\alpha c_\alpha(t)\,\phi_\alpha(\mathbf{r}) \tag{7.5}$$

where

$$\int d^3\mathbf{r}\,\phi_\alpha^*(\mathbf{r})\phi_\beta(\mathbf{r}) = \delta_{\alpha\beta}\,. \tag{7.6}$$

In this way, from the time-dependent Schrödinger equation we get

$$i\hbar\,\dot{c}_\alpha(t) = \epsilon_\alpha\,c_\alpha(t)\,, \tag{7.7}$$

whose general solution is obviously

$$c_\alpha(t) = c_\alpha(0)\,e^{-i\epsilon_\alpha t/\hbar}\,. \tag{7.8}$$

In this case the Schrödinger field $\psi(\mathbf{r}, t)$ is not a stationary state of the system, in fact

$$|\psi(\mathbf{r}, t)|^2 = \sum_{\alpha,\beta} c_\alpha^*(0)c_\beta(0)\,e^{-i(\epsilon_\alpha - \epsilon_\beta)t/\hbar}\,\phi_\alpha^*(\mathbf{r})\phi_\beta(\mathbf{r}) \tag{7.9}$$

depends on time t, while obviously its integral

$$\int d^3\mathbf{r}|\psi(\mathbf{r}, t)|^2 = \sum_\alpha |c_\alpha(0)|^2 \tag{7.10}$$

does not.

The constant of motion associated to the Schrödinger field $\psi(\mathbf{r}, t)$ is the average total energy of the system, given by

$$H = \int d^3\mathbf{r}\, \psi^*(\mathbf{r}, t) \left[-\frac{\hbar^2}{2m} \nabla^2 + U(\mathbf{r}) \right] \psi(\mathbf{r}, t) \,. \tag{7.11}$$

Inserting the expansion (7.5) into the total energy we find

$$H = \sum_\alpha \frac{\epsilon_\alpha}{2} \left(c_\alpha^* c_\alpha + c_\alpha c_\alpha^* \right) \,. \tag{7.12}$$

This energy is obviously independent on time: the time dependence of the complex amplitudes $c_\alpha^*(t)$ and $c_\alpha(t)$ cancels due to Eq. (7.8).

Instead of using the complex amplitudes $c_\alpha^*(t)$ and $c_\alpha(t)$ one can introduce the real variables

$$q_\alpha(t) = \sqrt{\frac{2\hbar}{\omega_\alpha}} \frac{1}{2} \left(c_\alpha(t) + c_\alpha^*(t) \right) \tag{7.13}$$

$$p_\alpha(t) = \sqrt{2\hbar\omega_\alpha} \frac{1}{2i} \left(c_\alpha(t) - c_\alpha^*(t) \right) \tag{7.14}$$

such that the energy of the matter field reads

$$H = \sum_\alpha \left(\frac{p_\alpha^2}{2} + \frac{1}{2} \omega_\alpha^2 q_\alpha^2 \right) \,, \tag{7.15}$$

where

$$\omega_\alpha = \frac{\epsilon_\alpha}{\hbar} \tag{7.16}$$

is the eigenfrequency associated to the eigenenergy ϵ_α.

This energy resembles that of infinitely many harmonic oscillators with unitary mass and frequency ω_α. It is written in terms of an infinite set of real harmonic oscillators: one oscillator for each mode characterized by quantum numbers α and frequency ω_α. It is clear the analogy between the energy of the Schrödinger field and the energy of the classical electromagnetic field.

7.2 Second Quantization of the Schrödinger Field

The canonical quantization of the classical Hamiltonian (7.15) is obtained by promoting the real coordinates q_α and the real momenta p_α to operators:

$$q_\alpha \rightarrow \hat{q}_\alpha \,, \tag{7.17}$$

$$p_\alpha \rightarrow \hat{p}_\alpha \,, \tag{7.18}$$

satisfying the commutation relations

$$[\hat{q}_\alpha, \hat{p}_\beta] = i\hbar\,\delta_{\alpha\beta}\;, \tag{7.19}$$

The quantum Hamiltonian is thus given by

$$\hat{H} = \sum_\alpha \left(\frac{\hat{p}_\alpha^2}{2} + \frac{1}{2}\omega_\alpha^2\hat{q}_\alpha^2 \right)\;. \tag{7.20}$$

The formal difference between Eqs. (7.15) and (7.20) is simply the presence of the "hat symbol" in the canonical variables.

We now introduce annihilation and creation operators

$$\hat{c}_\alpha = \sqrt{\frac{\omega_\alpha}{2\hbar}} \left(\hat{q}_\alpha + \frac{i}{\omega_\alpha}\,\hat{p}_\alpha \right)\;, \tag{7.21}$$

$$\hat{c}_\alpha^+ = \sqrt{\frac{\omega_\alpha}{2\hbar}} \left(\hat{q}_\alpha - \frac{i}{\omega_\alpha}\,\hat{p}_\alpha \right)\;, \tag{7.22}$$

which satisfy the commutation relations

$$[\hat{c}_\alpha, \hat{c}_\beta^+] = \delta_{\alpha,\beta}\;, \qquad [\hat{c}_\alpha, \hat{c}_\beta] = [\hat{c}_\alpha^+, \hat{c}_\beta^+] = 0\;, \tag{7.23}$$

and the quantum Hamiltonian (7.20) becomes

$$\hat{H} = \sum_\alpha \epsilon_\alpha \left(\hat{c}_\alpha^+\hat{c}_\alpha + \frac{1}{2} \right)\;. \tag{7.24}$$

Obviously this quantum Hamiltonian can be directly obtained from the classical one, given by Eq. (7.12), by promoting the complex amplitudes c_α and c_α^* to operators:

$$c_\alpha \to \hat{c}_\alpha\;, \tag{7.25}$$

$$c_\alpha^* \to \hat{c}_\alpha^+\;, \tag{7.26}$$

satisfying the commutation relations (7.23).

The operators $\hat{c}_\alpha$ and $\hat{c}_\alpha^+$ act in the Fock space of the "particles" of the Schrödinger field. A generic state of this Fock space is given by

$$| \ldots n_\alpha \ldots n_\beta \ldots n_\gamma \ldots \rangle\;, \tag{7.27}$$

meaning that there are n_α particles in the single-particle state $|\alpha\rangle$, n_β particles in the single-particle state $|\beta\rangle$, n_γ particles in the single-particle state $|\gamma\rangle$, et cetera. The operators $\hat{c}_\alpha$ and $\hat{c}_\alpha^+$ are called annihilation and creation operators because they respectively destroy and create one particle in the single-particle state $|\alpha\rangle$, namely

$$\hat{c}_\alpha | \dots n_\alpha \dots \rangle = \sqrt{n_\alpha} \, | \dots n_\alpha - 1 \dots \rangle \,, \tag{7.28}$$

$$\hat{c}_\alpha^+ | \dots n_\alpha \dots \rangle = \sqrt{n_\alpha + 1} \, | \dots n_\alpha + 1 \dots \rangle \,. \tag{7.29}$$

Note that these properties follow directly from the commutation relations (7.23). The vacuum state, where there are no particles, can be written as

$$|0\rangle = | \dots 0 \dots 0 \dots 0 \dots \rangle \,, \tag{7.30}$$

and

$$\hat{c}_\alpha |0\rangle = 0 \,, \tag{7.31}$$

$$\hat{c}_\alpha^+ |0\rangle = |1_\alpha\rangle = |\alpha\rangle \,, \tag{7.32}$$

where $|\alpha\rangle$ is such that

$$\langle \mathbf{r} | \alpha \rangle = \phi_\alpha(\mathbf{r}) \,. \tag{7.33}$$

From Eqs. (7.28) and (7.29) it follows immediately that

$$\hat{N}_\alpha = \hat{c}_\alpha^+ \hat{c}_\alpha \tag{7.34}$$

is the number operator which counts the number of particles in the single-particle state $|\alpha\rangle$, i.e.

$$\hat{N}_\alpha | \dots n_\alpha \dots \rangle = n_\alpha \, | \dots n_\alpha \dots \rangle \,. \tag{7.35}$$

Notice that the quantum Hamiltonian of the matter field can be written as

$$\hat{H} = \sum_\alpha \epsilon_\alpha \hat{N}_\alpha \,, \tag{7.36}$$

after removing the puzzling zero-point energy. This is the second-quantization Hamiltonian of non-interacting matter.

Introducing the time-evolution unitary operator

$$\hat{U}(t) = e^{-i\hat{H}t/\hbar} \,, \tag{7.37}$$

one finds that obviously

$$\hat{U}^+(t) \, \hat{c}_\alpha(0) \, \hat{U}(t) = \hat{c}_\alpha(0) \, e^{-i\epsilon_\alpha t/\hbar} \,. \tag{7.38}$$

In fact, one can write

$$i\hbar\frac{d}{dt}\left(\hat{U}^+(t)\,\hat{c}_\alpha(0)\,\hat{U}(t)\right) = i\hbar\frac{d}{dt}\left(e^{i\hat{H}t/\hbar}\,\hat{c}_\alpha(0)\,e^{-i\hat{H}t/\hbar}\right)$$

$$= i\hbar\left(\frac{i}{\hbar}\hat{H}\,\hat{c}_\alpha(t) - \frac{i}{\hbar}\hat{c}_\alpha(t)\,\hat{H}\right)$$

$$= \hat{c}_\alpha(t)\,\hat{H} - \hat{H}\hat{c}_\alpha(t)$$

$$= [\hat{c}_\alpha(t),\hat{H}]$$

$$= [\hat{c}_\alpha(t),\sum_\beta \epsilon_\beta\,\hat{c}_\beta^+(0)\,\hat{c}_\beta(0)]$$

$$= \sum_\beta \epsilon_\beta[\hat{c}_\alpha(t),\hat{c}_\beta^+(t)\,\hat{c}_\beta(t)]$$

$$= \sum_\beta \epsilon_\beta[\hat{c}_\alpha(t),\hat{c}_\beta^+(t)]\,\hat{c}_\beta(t)$$

$$= \sum_\beta \epsilon_\beta\,\delta_{\alpha,\beta}\,\hat{c}_\beta(t)$$

$$= \epsilon_\alpha\,\hat{c}_\alpha(t)\ . \tag{7.39}$$

Thus, one obtains

$$i\hbar\frac{d}{dt}\left(\hat{U}^+(t)\,\hat{c}_\alpha(0)\,\hat{U}(t)\right) = \epsilon_\alpha\,\hat{c}_\alpha(t)\ , \tag{7.40}$$

with solution

$$\hat{U}^+(t)\,\hat{c}_\alpha(0)\,\hat{U}(t) = \hat{c}_\alpha(0)\,e^{i\epsilon_\alpha t/\hbar}\ . \tag{7.41}$$

This is clearly the operatorial version of the classical solution (7.8), where

$$\hat{c}_\alpha(t) = \hat{U}^+(t)\,\hat{c}_\alpha(0)\,\hat{U}(t) \tag{7.42}$$

is the familiar expression of the time-evolution of the operator.

7.2.1 Bosonic and Fermionic Matter Field

The annihilation and creation operators $\hat{c}_\alpha$ and $\hat{c}_\alpha^+$ which satisfy the commutation rules (7.23) are called bosonic operators and the corresponding quantum field operator

$$\hat{\psi}(\mathbf{r},t) = \sum_\alpha \hat{c}_\alpha(t)\,\phi_\alpha(\mathbf{r}) \tag{7.43}$$

is the bosonic field operator. Indeed the commutation rules (7.23) imply Eqs. (7.28) and (7.29) and, as expected for bosons, there is no restriction on the number of particles n_α which can occupy in the single-particle state $|\alpha\rangle$.

To obtain fermionic properties it is sufficient to impose anti-commutation rules for the operators $\hat{c}_\alpha$ and $\hat{c}_\alpha^+$, i.e.

$$\{\hat{c}_\alpha, \hat{c}_\beta^+\} = \delta_{\alpha\beta} , \qquad \{\hat{c}_\alpha, \hat{c}_\beta\} = \{\hat{c}_\alpha^+, \hat{c}_\beta^+\} = 0 , \tag{7.44}$$

where $\{\hat{A}, \hat{B}\} = \hat{A}\hat{B} + \hat{B}\hat{A}$ are the anti-commutation brackets. An important consequence of anti-commutation is that

$$(\hat{c}_\alpha^+)^2 = 0 . \tag{7.45}$$

Notice that Eqs. (7.28) and (7.29) formally work also for fermions but with the constraints that $n_\alpha = 0$ or $n_\alpha = 1$ for any single-particle state $|\alpha\rangle$. In fact, for fermions

$$\hat{N}_\alpha^2 = \left(\hat{c}_\alpha^+ \hat{c}_\alpha\right)^2 = \hat{c}_\alpha^+ \hat{c}_\alpha \hat{c}_\alpha^+ \hat{c}_\alpha \tag{7.46}$$

$$= \hat{c}_\alpha^+ \left(1 - \hat{c}_\alpha^+ \hat{c}_\alpha\right) \hat{c}_\alpha = \hat{c}_\alpha^+ \hat{c}_\alpha \tag{7.47}$$

$$= \hat{N}_\alpha , \tag{7.48}$$

which implies that the eigenvalues of $\hat{N}_\alpha$ can be only 0 and 1. For a fermionic two-particle state we have

$$|1_\alpha 1_\beta\rangle = \hat{c}_\alpha^+ \hat{c}_\beta^+ |0\rangle = -\hat{c}_\beta^+ \hat{c}_\alpha^+ |0\rangle = -|1_\beta 1_\alpha\rangle , \tag{7.49}$$

i.e. the state is anti-symmetric under interchange of particle labels, and moreover

$$|2_\alpha\rangle = (\hat{c}_\alpha^+)^2 |0\rangle = 0 , \tag{7.50}$$

expressing the Pauli principle that two particles cannot be created in the same single-particle state. For fermions the field operator (7.43) is called fermionic field operator.

A remarkable property of the field operator $\hat{\psi}^+(\mathbf{r}, t)$, which works for bosons and fermions, is the following:

$$\hat{\psi}^+(\mathbf{r}, t)|0\rangle = |\mathbf{r}, t\rangle \tag{7.51}$$

that is the operator $\hat{\psi}^+(\mathbf{r}, t)$ creates a particle in the state $|\mathbf{r}, t\rangle$ from the vacuum state $|0\rangle$. In fact,

$$\hat{\psi}^+(\mathbf{r}, t)|0\rangle = \sum_\alpha \hat{c}_\alpha^+(t) \, \phi_\alpha^*(\mathbf{r})|0\rangle = \sum_\alpha \hat{c}_\alpha^+ e^{i\epsilon_\alpha t/\hbar} \langle \alpha | \mathbf{r}\rangle |0\rangle$$

$$= \sum_\alpha e^{i\epsilon_\alpha t/\hbar} |\alpha\rangle \langle \alpha | \mathbf{r}\rangle = |\mathbf{r}, t\rangle , \tag{7.52}$$

because

$$|\mathbf{r}, t\rangle = e^{i\hat{H}t/\hbar}|\mathbf{r}\rangle = \sum_\alpha |\alpha\rangle\langle\alpha|e^{i\hat{H}t/\hbar}\sum_\beta |\beta\rangle\langle\beta|\mathbf{r}\rangle = \sum_\alpha e^{i\epsilon_\alpha t/\hbar}|\alpha\rangle\langle\alpha|\mathbf{r}\rangle \quad (7.53)$$

with $\hat{c}_\alpha^+ = \hat{c}_\alpha^+(0)$, $\langle\alpha|\beta\rangle = \delta_{\alpha\beta}$, and $\sum_\alpha |\alpha\rangle\langle\alpha| = 1$.

It is straightforward to show that the bosonic field operator satisfies the following equal-time commutation rules

$$[\hat{\psi}(\mathbf{r}, t), \hat{\psi}^+(\mathbf{r}', t)] = \delta(\mathbf{r} - \mathbf{r}') , \quad (7.54)$$

while for the fermionic field operator one gets

$$\{\hat{\psi}(\mathbf{r}, t), \hat{\psi}^+(\mathbf{r}', t)\} = \delta(\mathbf{r} - \mathbf{r}') . \quad (7.55)$$

Let us prove Eq. (7.54). By using the expansion of the field operators one finds

$$\begin{aligned}
[\hat{\psi}(\mathbf{r}, t), \hat{\psi}^+(\mathbf{r}', t)] &= \sum_{\alpha,\beta} \phi_\alpha(\mathbf{r})\,\phi_\beta^*(\mathbf{r}')\,[c_\alpha, c_\beta^+] \\
&= \sum_{\alpha,\beta} \phi_\alpha(\mathbf{r})\,\phi_\beta^*(\mathbf{r}')\,\delta_{\alpha,\beta} = \sum_\alpha \phi_\alpha(\mathbf{r})\,\phi_\alpha^*(\mathbf{r}') \\
&= \sum_\alpha \langle\mathbf{r}|\alpha\rangle\,\langle\alpha|\mathbf{r}'\rangle = \langle\mathbf{r}|\sum_\alpha |\alpha\rangle\,\langle\alpha|\mathbf{r}'\rangle \\
&= \langle\mathbf{r}|\mathbf{r}'\rangle = \delta(\mathbf{r} - \mathbf{r}') .
\end{aligned} \quad (7.56)$$

The proof for fermions is practically the same, with anti-commutators instead of commutators.

Finally, we observe that the many-body quantum Hamiltonian (7.36) can be written, both for bosons and fermions, in the elegant form

$$\hat{H} = \int d^3\mathbf{r}\,\hat{\psi}^+(\mathbf{r}, t)\left[-\frac{\hbar^2}{2m}\nabla^2 + U(\mathbf{r})\right]\hat{\psi}(\mathbf{r}, t) . \quad (7.57)$$

This quantum Hamiltonian can be directly obtained from the classical one, given by Eq. (7.11), by promoting the complex classical field $\psi(\mathbf{r}, t)$ and $\hat{\psi}(\mathbf{r}, t)$ to quantum field operators:

$$\psi(\mathbf{r}, t) \rightarrow \hat{\psi}(\mathbf{r}, t) , \quad (7.58)$$
$$\psi^*(\mathbf{r}, t) \rightarrow \hat{\psi}^+(\mathbf{r}, t) , \quad (7.59)$$

satisfying the commutation relations (7.54) of bosons or the anti-commutation relations (7.55) of fermions.

7.3 Connection Between First and Second Quantization

In this section we shall analyze relationships between the formalism of second quantization, where the number of particles is apriori not fixed, and the formalism of first quantization, where the number of particles is instead fixed.

First we observe that, within the formalism of second quantization of field operators, the time-independent one-body density operator is defined as

$$\hat{\rho}(\mathbf{r}) = \hat{\psi}^+(\mathbf{r})\,\hat{\psi}(\mathbf{r})\,, \tag{7.60}$$

and it is such that

$$\hat{N} = \int d^3\mathbf{r}\,\hat{\rho}(\mathbf{r}) \tag{7.61}$$

is the total number operator. By using the expansion

$$\hat{\psi}(\mathbf{r}) = \sum_\alpha \hat{c}_\alpha\,\phi_\alpha(\mathbf{r}) \tag{7.62}$$

one finds immediately

$$\hat{\rho}(\mathbf{r}) = \sum_{\alpha,\beta} \hat{c}_\alpha^+ \hat{c}_\beta\,\phi_\alpha^*(\mathbf{r})\,\phi_\beta(\mathbf{r}) \tag{7.63}$$

and also

$$\hat{N} = \sum_\alpha \hat{c}_\alpha^+ \hat{c}_\alpha = \sum_\alpha \hat{N}_\alpha\,, \tag{7.64}$$

because we consider an orthonormal basis of single-particle wavefunctions $\phi_\alpha(\mathbf{r})$.

A remarkable connection between second quantization and first quantization is made explicit by the formula

$$\hat{\rho}(\mathbf{r})\,|\mathbf{r}_1\mathbf{r}_2\ldots\mathbf{r}_N\rangle = \sum_{i=1}^{N} \delta(\mathbf{r}-\mathbf{r}_i)\,|\mathbf{r}_1\mathbf{r}_2\ldots\mathbf{r}_N\rangle\,, \tag{7.65}$$

where $\rho(\mathbf{r}) = \sum_{i=1}^{N} \delta(\mathbf{r}-\mathbf{r}_i)$ is the one-body density function introduced in the previous chapter. This formula can be proved, for bosons, as follows:

$$
\begin{aligned}
\hat{\rho}(\mathbf{r})\,|\mathbf{r}_1\mathbf{r}_2\ldots\mathbf{r}_N\rangle &= \hat{\psi}^+(\mathbf{r})\,\hat{\psi}(\mathbf{r})\,|\mathbf{r}_1\mathbf{r}_2\ldots\mathbf{r}_N\rangle \\
&= \hat{\psi}^+(\mathbf{r})\,\hat{\psi}(\mathbf{r})\,\hat{\psi}^+(\mathbf{r}_1)\,\hat{\psi}^+(\mathbf{r}_2)\ldots\hat{\psi}^+(\mathbf{r}_N)\,|0\rangle \\
&= \hat{\psi}^+(\mathbf{r})\left(\delta(\mathbf{r}-\mathbf{r}_1) + \hat{\psi}^+(\mathbf{r}_1)\,\hat{\psi}(\mathbf{r})\right)\hat{\psi}^+(\mathbf{r}_2)\ldots\hat{\psi}^+(\mathbf{r}_N)\,|0\rangle \\
&= \delta(\mathbf{r}-\mathbf{r}_1)\,|\mathbf{r}_1\mathbf{r}_2\ldots\mathbf{r}_N\rangle \\
&\quad + \hat{\psi}^+(\mathbf{r})\,\hat{\psi}^+(\mathbf{r}_1)\,\hat{\psi}(\mathbf{r})\,\hat{\psi}^+(\mathbf{r}_2)\ldots\hat{\psi}^+(\mathbf{r}_N)\,|0\rangle \\
&= \delta(\mathbf{r}-\mathbf{r}_1)\,|\mathbf{r}_1\mathbf{r}_2\ldots\mathbf{r}_N\rangle
\end{aligned}
$$

$$+ \hat{\psi}^+(\mathbf{r})\, \hat{\psi}^+(\mathbf{r}_1) \left(\delta(\mathbf{r} - \mathbf{r}_2) + \hat{\psi}^+(\mathbf{r}_2)\, \hat{\psi}(\mathbf{r}) \right) \hat{\psi}^+(\mathbf{r}_3) \ldots \hat{\psi}^+(\mathbf{r}_N) \, |0\rangle$$

$$= \delta(\mathbf{r} - \mathbf{r}_1)\, |\mathbf{r}_1 \mathbf{r}_2 \ldots \mathbf{r}_N\rangle + \delta(\mathbf{r} - \mathbf{r}_2)\, |\mathbf{r}_1 \mathbf{r}_2 \ldots \mathbf{r}_N\rangle$$

$$+ \hat{\psi}^+(\mathbf{r})\, \hat{\psi}^+(\mathbf{r}_2)\, \hat{\psi}(\mathbf{r})\, \hat{\psi}^+(\mathbf{r}_3) \ldots \hat{\psi}^+(\mathbf{r}_N) \, |0\rangle$$

$$= \ldots$$

$$= \sum_{i=1}^{N} \delta(\mathbf{r} - \mathbf{r}_i)\, |\mathbf{r}_1 \mathbf{r}_2 \ldots \mathbf{r}_N\rangle$$

$$+ \hat{\psi}^+(\mathbf{r})\, \hat{\psi}^+(\mathbf{r}_2)\, \hat{\psi}^+(\mathbf{r}_3) \ldots \hat{\psi}^+(\mathbf{r}_N)\, \hat{\psi}(\mathbf{r})\, |0\rangle$$

$$= \sum_{i=1}^{N} \delta(\mathbf{r} - \mathbf{r}_i)\, |\mathbf{r}_1 \mathbf{r}_2 \ldots \mathbf{r}_N\rangle \, , \tag{7.66}$$

taking into account the commutation relations of the field operators $\hat{\psi}(\mathbf{r})$ and $\hat{\psi}^+(\mathbf{r})$. For fermions the proof proceeds in a very similar way.

We can further investigate the connection between second quantization and first quantization by analyzing the Hamiltonian operator. In first quantization, the non-relativistic quantum Hamiltonian of N interacting identical particles in the external potential $U(\mathbf{r})$ is given by

$$\hat{H}^{(N)} = \sum_{i=1}^{N} \left[-\frac{\hbar^2}{2m} \nabla_i^2 + U(\mathbf{r}_i) \right] + \frac{1}{2} \sum_{\substack{i,j=1 \\ i \neq j}}^{N} V(\mathbf{r}_i - \mathbf{r}_j) = \sum_{i=1}^{N} \hat{h}_i + \frac{1}{2} \sum_{\substack{i,j=1 \\ i \neq j}}^{N} V_{ij} \, , \tag{7.67}$$

where $V(\mathbf{r} - \mathbf{r}')$ is the inter-particle potential. In second quantization, the quantum field operator can be written as

$$\hat{\psi}(\mathbf{r}) = \sum_{\alpha} \hat{c}_{\alpha}\, \phi_{\alpha}(\mathbf{r}) \tag{7.68}$$

where the $\phi_{\alpha}(\mathbf{r}) = \langle \mathbf{r} | \alpha \rangle$ are the eigenfunctions of $\hat{h}$ such that $\hat{h} | \alpha \rangle = \epsilon_{\alpha} | \alpha \rangle$, and $\hat{c}_{\alpha}$ and $\hat{c}_{\alpha}^+$ are the annihilation and creation operators of the single-particle state $| \alpha \rangle$. We now introduce the quantum many-body Hamiltonian

$$\hat{H} = \sum_{\alpha} \epsilon_{\alpha}\, \hat{c}_{\alpha}^+ \hat{c}_{\alpha} + \sum_{\alpha\beta\gamma\delta} V_{\alpha\beta\gamma\delta}\, \hat{c}_{\alpha}^+ \hat{c}_{\beta}^+ \hat{c}_{\delta} \hat{c}_{\gamma} \, , \tag{7.69}$$

where

$$V_{\alpha\beta\delta\gamma} = \int d^3\mathbf{r}\, d^3\mathbf{r}'\, \phi_{\alpha}^*(\mathbf{r})\, \phi_{\beta}^*(\mathbf{r}')\, V(\mathbf{r} - \mathbf{r}')\, \phi_{\delta}(\mathbf{r}')\, \phi_{\gamma}(\mathbf{r}) \, . \tag{7.70}$$

This Hamiltonian can be also written as

$$\hat{H} = \int d^3\mathbf{r} \, \hat{\psi}^+(\mathbf{r}) \left[-\frac{\hbar^2}{2m} \nabla^2 + U(\mathbf{r}) \right] \hat{\psi}(\mathbf{r})$$
$$+ \frac{1}{2} \int d^3\mathbf{r} \, d^3\mathbf{r}' \, \hat{\psi}^+(\mathbf{r}) \, \hat{\psi}^+(\mathbf{r}') \, V(\mathbf{r} - \mathbf{r}') \, \hat{\psi}(\mathbf{r}') \, \hat{\psi}(\mathbf{r}) \, . \tag{7.71}$$

We are now ready to show the meaningful connection between the second-quantization Hamiltonian $\hat{H}$ and the first-quantization Hamiltonian $\hat{H}^{(N)}$, which is given by the formula

$$\hat{H} | \mathbf{r}_1 \mathbf{r}_2 \dots \mathbf{r}_N \rangle = \hat{H}^{(N)} | \mathbf{r}_1 \mathbf{r}_2 \dots \mathbf{r}_N \rangle \, . \tag{7.72}$$

This formula can be proved following the same path of Eq. (7.66). In fact, one finds that

$$\hat{\psi}^+(\mathbf{r}) \, \hat{h}(\mathbf{r}) \, \hat{\psi}(\mathbf{r}) \, | \mathbf{r}_1 \mathbf{r}_2 \dots \mathbf{r}_N \rangle = \sum_{i=1}^{N} \hat{h}(\mathbf{r}_i) \delta(\mathbf{r} - \mathbf{r}_i) \, | \mathbf{r}_1 \mathbf{r}_2 \dots \mathbf{r}_N \rangle \tag{7.73}$$

and also

$$\hat{\psi}^+(\mathbf{r}) \, \hat{\psi}^+(\mathbf{r}') \, V(\mathbf{r}, \mathbf{r}') \, \hat{\psi}(\mathbf{r}') \, \hat{\psi}(\mathbf{r}) \, | \mathbf{r}_1 \mathbf{r}_2 \dots \mathbf{r}_N \rangle = \sum_{\substack{i,j=1 \\ i \neq j}}^{N} V(\mathbf{r}_i, \mathbf{r}_i) \delta(\mathbf{r} - \mathbf{r}_i) \, \delta(\mathbf{r}' - \mathbf{r}_j) \, | \mathbf{r}_1 \mathbf{r}_2 \dots \mathbf{r}_N \rangle \, .$$
$$\tag{7.74}$$

From these two expressions Eq. (7.72) follows immediately, after space integration. Similarly to Eqs. (7.65), (7.72) displays a deep connection between second and first quantization.

Up to now in this section we have considered time-independent quantum field operators. The time-dependent equation of motion for the field operator $\hat{\psi}(\mathbf{r}, t)$ is easily obtained from the Hamiltonian (7.71) by using the familiar Heisenberg equation

$$i\hbar \frac{\partial}{\partial t} \hat{\psi} = [\hat{\psi}, \hat{H}] \, , \tag{7.75}$$

which gives

$$i\hbar \frac{\partial}{\partial t} \hat{\psi} = \left[-\frac{\hbar^2}{2m} \nabla^2 + U(\mathbf{r}) \right] \hat{\psi}(\mathbf{r}) + \int d^3\mathbf{r}' \, \hat{\psi}^+(\mathbf{r}') \, V(\mathbf{r} - \mathbf{r}') \, \hat{\psi}(\mathbf{r}') \, \hat{\psi}(\mathbf{r}) \, . \tag{7.76}$$

This exact equation for the field operator $\hat{\psi}(\mathbf{r}, t)$ is second-quantized version of the Hartree-like time-dependent nonlinear Schrödinger equation of a complex wavefunction (classical field) $\psi(\mathbf{r}, t)$. Note that, in practice, one can formally use the simpler formula

$$i\hbar \frac{\partial}{\partial t} \hat{\psi} = \frac{\delta \hat{H}}{\delta \hat{\psi}^+} \tag{7.77}$$

to determine the Heisenberg equation of motion, where

$$\frac{\delta \hat{H}}{\delta \hat{\psi}+} = \left(\frac{\delta H}{\delta \psi^*}\right)_{\psi=\hat{\psi},\psi^*=\hat{\psi}+} . \tag{7.78}$$

Obviously, the equation of motion for the time-dependent field operator $\hat{\psi}^+(\mathbf{r}, t)$ is obtained in the same way.

Similarly, the time-dependent equation of motion of the field operator $\hat{c}_\alpha(t)$ is obtained from the Hamiltonian (7.70) by using the Heisenberg equation

$$i\hbar \frac{d}{dt} \hat{c}_\alpha = [\hat{c}_\alpha, \hat{H}] , \tag{7.79}$$

which gives

$$i\hbar \frac{d}{dt} \hat{c}_\alpha = \hat{c}_\alpha + \sum_{\beta\gamma\delta} V_{\alpha\beta\gamma\delta} \, \hat{c}_\beta^+ \hat{c}_\delta \hat{c}_\gamma . \tag{7.80}$$

Again, one can formally can the simpler formula

$$i\hbar \frac{d}{dt} \hat{c}_\alpha = \frac{\partial \hat{H}}{\partial \hat{c}_\alpha^+} \tag{7.81}$$

to determine the Heisenberg equation of motion of $\hat{c}_\alpha$, where

$$\frac{\partial \hat{H}}{\partial \hat{a}_\alpha^+} = \left(\frac{\partial H}{\partial a_\alpha^*}\right)_{a_\alpha=\hat{a}_\alpha,a_\alpha^*=\hat{a}_\alpha^+} . \tag{7.82}$$

7.4 Coherent States for Bosonic and Fermionic Matter Fields

Nowadays it is possible to produce dilute and ultracold bosonic gases made of alkali-metal atoms in a confining magnetic or optical potential $U(\mathbf{r})$. For instance, one million of ^{87}Rb atoms at the temperature of 100 nK in the lowest single-particle state. The system can be so dilute that the inter-atomic interaction can be neglected. Under these conditions one has a pure Bose-Einstein condensate made of non-interacting bosons described by the Hamiltonian

$$\hat{H} = \int d^3\mathbf{r} \, \hat{\psi}^+(\mathbf{r}) \left[-\frac{\hbar^2}{2m}\nabla^2 + U(\mathbf{r})\right] \hat{\psi}(\mathbf{r}) = \sum_\alpha \epsilon_\alpha \, \hat{c}_\alpha^+ \hat{c}_\alpha . \tag{7.83}$$

If there are exactly N bosons in the lowest single-particle state of energy ϵ_0, the ground-state of the system is a Fock state given by

$$|F_N\rangle = \frac{1}{\sqrt{N!}} \left(\hat{c}_0^+\right)^N |0\rangle = |N, 0, 0, \dots\rangle . \tag{7.84}$$

It is then straightforward to show that

$$\langle F_N|\hat{\psi}(\mathbf{r})|F_N\rangle = 0 , \tag{7.85}$$

while the expectation value of $\hat{\psi}^+(\mathbf{r})\hat{\psi}(\mathbf{r})$ is given by

$$\langle F_N|\hat{\psi}^+(\mathbf{r})\hat{\psi}(\mathbf{r})|F_N\rangle = N\,|\psi_0(\mathbf{r})|^2 . \tag{7.86}$$

These results are the exact analog of what we have seen for the light field in Chap. 2. In the spirit of quantum optics one can suppose that the number of massive bosons in the matter field is not fixed, in other words that the system is not in a Fock state. We then introduce the coherent state $|c_0\rangle$, such that

$$\hat{c}_0|c_0\rangle = c_0|c_0\rangle , \tag{7.87}$$

with

$$\langle c_0|c_0\rangle = 1 . \tag{7.88}$$

The coherent state $|c_0\rangle$ is thus the eigenstate of the annihilation operator $\hat{c}_0$ with complex eigenvalue $c_0 = |c_0|e^{i\theta_0}$. $|c_0\rangle$ does not have a fixed number of bosons, i.e. it is not an eigenstate of the number operator $\hat{N}$ nor of $\hat{N}_0$, and it is not difficult to show that $|c_0\rangle$ can be expanded in terms of number (Fock) states $|N, 0, 0, \dots\rangle$ as follows

$$|c_0\rangle = e^{-|c_0|^2/2} \sum_{N=0}^{\infty} \frac{c_0^N}{\sqrt{N!}} |N, 0, 0, \dots\rangle . \tag{7.89}$$

From Eq. (7.87) one immediately finds

$$\bar{N} = \langle c_0|\hat{N}|c_0\rangle = |c_0|^2 , \tag{7.90}$$

and it is natural to set

$$c_0 = \sqrt{\bar{N}}\,e^{i\theta_0} , \tag{7.91}$$

where $\bar{N}$ is the average number of bosons in the coherent state, while θ_0 is the phase of the coherent state. The expectation value of the matter field $\hat{\psi}(\mathbf{r})$ in the coherent state $|c_0\rangle$ reads

$$\langle c_0|\hat{\psi}(\mathbf{r})|c_0\rangle = \sqrt{\bar{N}}\,e^{i\theta_0}\,\phi_0(\mathbf{r}) , \tag{7.92}$$

with $\omega_0 = \epsilon_0/\hbar$, while the expectation value of $\hat{\psi}^+(\mathbf{r})\hat{\psi}(\mathbf{r})$ is given by

$$\langle c_0 | \hat{\psi}^+(\mathbf{r})\hat{\psi}(\mathbf{r}) | c_0 \rangle = \bar{N} \, |\phi_0(\mathbf{r})|^2 \,. \tag{7.93}$$

Up to now we have considered coherent states only for bosonic fields: the electromagnetic field and the bosonic matter field. Are there coherent states also for the fermionic field? The answer is indeed positive.

For simplicity, let us consider a single mode of the fermionic field described by the anti-commuting fermionic operators $\hat{c}$ and $\hat{c}^+$, such that

$$\hat{c}\hat{c}^+ + \hat{c}^+\hat{c} = 1 \,, \qquad \hat{c}^2 = (\hat{c}^+)^2 = 0 \,. \tag{7.94}$$

The fermionic coherent state $|c\rangle$ is defined as the eigenstate of the fermionic annihilation operator $\hat{c}$, namely

$$\hat{c}|c\rangle = c|c\rangle \,, \tag{7.95}$$

where c is the corresponding eigenvalue. It is immediate to verify that, for mathematical consistency, this eigenvalue c must satisfy the following relationships

$$c\bar{c} + \bar{c}c = 1 \,, \qquad c^2 = \bar{c}^2 = 0 \,, \tag{7.96}$$

where $\bar{c}$ is such that

$$\langle c|\hat{c}^+ = \bar{c}\langle c| \,. \tag{7.97}$$

Obviously c and $\bar{c}$ are not complex numbers. They are instead Grassmann numbers, namely elements of the Grassmann linear algebra $\{1, c, \bar{c}, \bar{c}c\}$ characterized by the independent basis elements $1, c, \bar{c}$, with 1 the identity (neutral) element. The most general function on this Grassmann algebra is given by

$$f(\bar{c}, c) = f_{11} + f_{12}\,c + f_{21}\,\bar{c} + f_{22}\,\bar{c}c \,, \tag{7.98}$$

where $f_{11}, f_{12}, f_{21}, f_{22}$ are complex numbers. In fact, the function $f(\bar{c}, c)$ does not have higher powers of $c, \bar{c}$ and $\bar{c}c$ because they are identically zero. The differentiation with respect to the Grassmann variable c is defined as

$$\frac{\partial}{\partial c} f(\bar{c}, c) = f_{12} - f_{22}\,\bar{c} \,, \tag{7.99}$$

where the minus sign occurs because one need to permute $\bar{c}$ and c before differentiation. This is called left differentiation. Clearly, one has also

$$\frac{\partial}{\partial \bar{c}} f(\bar{c}, c) = f_{21} + f_{22}\,c \tag{7.100}$$

and

$$\frac{\partial^2}{\partial\bar{c}\partial c}f(\bar{c},c) = -\frac{\partial^2}{\partial c\partial\bar{c}}f(\bar{c},c) = -f_{22}\ . \tag{7.101}$$

The integration over Grassmann numbers is a bit more tricky. It is defined as equivalent to differentiation, i.e.

$$\int dc = \frac{\partial}{\partial c}\ ,\quad \int d\bar{c} = \frac{\partial}{\partial\bar{c}}\ ,\quad \int d\bar{c}\,dc = \frac{\partial^2}{\partial\bar{c}\partial c}\ ,\quad \int dc\,d\bar{c} = \frac{\partial^2}{\partial c\partial\bar{c}}\ . \tag{7.102}$$

These integrals are not integrals in the Lebesgue sense. They are called integrals simply because they have some properties of Lebesgue integrals: the linearity and the fundamental property of ordinary integrals over functions vanishing at infinity that the integral of an exact differential form is zero. Notice that the mathematical theory of this kind of integral (now called Berezin integral) with anticommuting Grassmann variables was invented and developed by Felix Berezin in 1966. The Berezin integral is mainly used for the path integral formulation of quantum field theory. A discussion of the path integral formulation of quantum field theory, also called functional integration, is outside the scope of this book.

To conclude this section we stress that, in full generality, the classical analog of the bosonic quantum field operator

$$\hat{\psi}(\mathbf{r}) = \sum_j \phi_j(\mathbf{r})\,\hat{c}_j \tag{7.103}$$

is the complex classical field

$$\psi(\mathbf{r}) = \sum_j \phi_\alpha(\mathbf{r})\,c_j \tag{7.104}$$

such that

$$\hat{\psi}(\mathbf{r})|\psi\rangle = \psi(\mathbf{r})|\psi\rangle\ , \tag{7.105}$$

where

$$|\psi\rangle = \prod_j |c_j\rangle \tag{7.106}$$

is the bosonic coherent state of the system, $|c_j\rangle$ is the coherent state of the bosonic operator $\hat{c}_j$, and c_j is its complex eigenvalue. Similarly, one can introduce the pseudo-classical Grassmann analog of the fermionic field operator by using fermionic coherent states.

7.5 Quantum Matter Field at Finite Temperature

Let us consider the non-interacting matter field in thermal equilibrium with a bath at the temperature T. The relevant quantity to calculate all thermodynamical properties of the system is the grand-canonical partition function $\mathcal{Z}$, given by

$$\mathcal{Z} = Tr[e^{-\beta(\hat{H}-\mu\hat{N})}] \tag{7.107}$$

where $\beta = 1/(k_B T)$ with k_B the Boltzmann constant,

$$\hat{H} = \sum_\alpha \epsilon_\alpha \hat{N}_\alpha , \tag{7.108}$$

is the quantum Hamiltonian,

$$\hat{N} = \sum_\alpha \hat{N}_\alpha \tag{7.109}$$

is total number operator, and μ is the chemical potential, fixed by the conservation of the average particle number. This implies that

$$\mathcal{Z} = \sum_{\{n_\alpha\}} \langle \dots n_\alpha \dots | e^{-\beta(\hat{H}-\mu\hat{N})} | \dots n_\alpha \dots \rangle = \sum_{\{n_\alpha\}} \langle \dots n_\alpha \dots | e^{-\beta\sum_\alpha(\epsilon_\alpha-\mu)\hat{N}_\alpha} | \dots n_\alpha \dots \rangle$$

$$= \sum_{\{n_\alpha\}} e^{-\beta\sum_\alpha(\epsilon_\alpha-\mu)n_\alpha} = \sum_{\{n_\alpha\}} \prod_\alpha e^{-\beta(\epsilon_\alpha-\mu)n_\alpha} = \prod_\alpha \sum_{n_\alpha} e^{-\beta(\epsilon_\alpha-\mu)n_\alpha}$$

$$= \prod_\alpha \sum_{n=0}^\infty e^{-\beta(\epsilon_\alpha-\mu)n} = \prod_\alpha \frac{1}{1-e^{-\beta(\epsilon_\alpha-\mu)}} \qquad \text{for bosons} \tag{7.110}$$

$$= \prod_\alpha \sum_{n=0}^1 e^{-\beta(\epsilon_\alpha-\mu)n} = \prod_\alpha \left(1+e^{-\beta(\epsilon_\alpha-\mu)}\right) \qquad \text{for fermions} \tag{7.111}$$

Quantum statistical mechanics dictates that the thermal average of any operator $\hat{A}$ is obtained as

$$\langle\hat{A}\rangle_T = \frac{1}{\mathcal{Z}} Tr[\hat{A}\, e^{-\beta(\hat{H}-\mu\hat{N})}] . \tag{7.112}$$

Let us suppose that $\hat{A} = \hat{H}$, it is then quite easy to show that

$$\langle\hat{H}'\rangle_T = \frac{1}{\mathcal{Z}} Tr[(\hat{H}-\mu\hat{N})\, e^{-\beta(\hat{H}-\mu\hat{N})}] = -\frac{\partial}{\partial\beta}\ln\left(Tr[e^{-\beta(\hat{H}-\mu\hat{N})}]\right) = -\frac{\partial}{\partial\beta}\ln(\mathcal{Z}) . \tag{7.113}$$

By using Eq. (7.110) or Eq. (7.111) we immediately obtain

$$\ln(\mathcal{Z}) = \mp\sum_\alpha \ln\left(1\mp e^{-\beta(\epsilon_\alpha-\mu)}\right) , \tag{7.114}$$

where $-$ is for bosons and $+$ for fermions, and finally from Eq. (7.113) we get

$$\langle \hat{H} \rangle_T = \sum_\alpha \epsilon_\alpha \langle \hat{N}_\alpha \rangle_T \ , \tag{7.115}$$

with

$$\langle \hat{N} \rangle_T = \sum_\alpha \frac{1}{e^{\beta(\epsilon_\alpha - \mu)} \mp 1} \ . \tag{7.116}$$

Notice that the zero-temperature limit, i.e. $\beta \to \infty$, for fermions gives

$$\langle \hat{H} \rangle_0 = \sum_\alpha \epsilon_\alpha \langle \hat{N}_\alpha \rangle_0 \tag{7.117}$$

with

$$\langle \hat{N} \rangle_0 = \sum_\alpha \Theta (\mu - \epsilon_\alpha) \ , \tag{7.118}$$

where the chemical potential μ at zero temperature is nothing but the Fermi energy ϵ_F, i.e. $\epsilon_F = \mu(T = 0)$.

7.6 Matter-Radiation Interaction

In Chap. 3 we have discussed the quantum electrodynamics by using first quantization for the matter and second quantization of the electromagnetic radiation. Here we write down the fully second-quantized Hamiltonian. It is given by

$$\hat{H} = \hat{H}_{matt} + \hat{H}_{rad} + \hat{H}_I \tag{7.119}$$

where

$$\hat{H}_{matt} = \int d^3\mathbf{r} \, \hat{\psi}^+(\mathbf{r}, t) \left[-\frac{\hbar^2}{2m} \nabla^2 + U(\mathbf{r}) \right] \hat{\psi}(\mathbf{r}, t) \ , \tag{7.120}$$

$$\hat{H}_{rad} = \int d^3\mathbf{r} \left(\frac{\varepsilon_0}{2} \left(\frac{\partial \hat{\mathbf{A}}(\mathbf{r}, t)}{\partial t} \right)^2 + \frac{1}{2\mu_0} \left(\nabla \wedge \hat{\mathbf{A}}(\mathbf{r}, t) \right)^2 \right) \ , \tag{7.121}$$

$$\hat{H}_I = \int d^3\mathbf{r} \, \hat{\psi}^+(\mathbf{r}, t) \left[-\frac{e}{m} \hat{\mathbf{A}}(\mathbf{r}, t) \cdot (-i\hbar\nabla) + \frac{e^2}{2m} \hat{\mathbf{A}}(\mathbf{r}, t)^2 \right] \hat{\psi}(\mathbf{r}, t) \ . \tag{7.122}$$

with $\hat{\psi}(\mathbf{r}, t)$ the scalar field operator of electrons in the external potential $U(\mathbf{r})$, which includes the instantaneous electrostatic Coulomb potential, and $\hat{\mathbf{A}}(\mathbf{r}, t)$ the vector field operator of the electromagnetic radiation. Notice that the coupling Hamiltonian $\hat{H}_I$ describes the very specific situation of the charged matter (usually the electron gas) coupled to radiation, which is an important but by no means unique case of matter interacting with electromagnetic radiation using second quantization methods. One often uses similar techniques to describe e.g. neutral atomic gases as we shall show in the last section.

Within the dipole approximation, where one neglects the quadratic term of the vector potential and moreover one assumes that the spatial behavior of the vector potential field changes more slowly than the matter field, the interaction Hamiltonian becomes

$$\hat{H}_I \simeq \hat{H}_D = \int d^3\mathbf{r}\, \hat{\psi}^+(\mathbf{r}, t) \left[-\frac{e}{m} \hat{\mathbf{A}}(\mathbf{0}, t) \cdot (-i\hbar\nabla) \right] \hat{\psi}(\mathbf{r}, t) \; . \tag{7.123}$$

By performing the following expansion for the matter field

$$\hat{\psi}(\mathbf{r}, t) = \sum_\alpha \hat{c}_\alpha\, e^{-i\epsilon_\alpha t/\hbar}\, \phi_\alpha(\mathbf{r}) \; , \tag{7.124}$$

where the $\phi_\alpha(\mathbf{r}) = \langle \mathbf{r} | \alpha \rangle$ are the eigenfunctions of $\hat{h} = -\frac{\hbar^2}{2m}\nabla^2 + U(\mathbf{r})$, and the following expansion for the radiation field

$$\hat{\mathbf{A}}(\mathbf{r}, t) = \sum_{ks} \sqrt{\frac{\hbar}{2\epsilon_0 \omega_k V}} \left[\hat{a}_{ks}\, e^{i(\mathbf{k}\cdot\mathbf{r} - \omega_k t)} + \hat{a}^+_{ks}\, e^{-i(\mathbf{k}\cdot\mathbf{r} - \omega_k t)} \right] \varepsilon_{ks} \; , \tag{7.125}$$

we finally obtain

$$\hat{H}_{matt} = \sum_\alpha \epsilon_\alpha\, \hat{c}^+_\alpha \hat{c}_\alpha \; , \tag{7.126}$$

$$\hat{H}_{rad} = \sum_{ks} \hbar\omega_k\, \hat{a}^+_{ks} \hat{a}_{ks} \; , \tag{7.127}$$

$$\hat{H}_D = \hbar \sum_{\alpha\beta ks} g_{\alpha\beta ks}\, \hat{c}^+_\beta \hat{c}_\alpha \left(\hat{a}_{ks}\, e^{-i\omega_k t} + \hat{a}^+_{ks}\, e^{i\omega_k t} \right) e^{i(\epsilon_\beta - \epsilon_\alpha)t/\hbar} \tag{7.128}$$

where

$$g_{\alpha\beta ks} = -\frac{e}{m} \sqrt{\frac{1}{2\epsilon_0 \hbar\omega_k V}}\; \varepsilon_{ks} \cdot \int d^3\mathbf{r}\, \phi^*_\beta(\mathbf{r})(-i\hbar\nabla)\phi_\alpha(\mathbf{r}) \; . \tag{7.129}$$

Notice that

$$\int d^3\mathbf{r}\, \phi^*_\beta(\mathbf{r})(-i\hbar\nabla)\phi_\alpha(\mathbf{r}) = \langle \beta | \hat{\mathbf{p}} | \alpha \rangle = i\, m\, \omega_{\beta\alpha} \langle \beta | \mathbf{r} | \alpha \rangle \; , \tag{7.130}$$

where $\omega_{\beta\alpha} = (\epsilon_\beta - \epsilon_\alpha)/\hbar$ and consequently we can write

$$\hat{H}_D = \hbar \sum_{\alpha\beta\mathbf{ks}} g_{\alpha\beta\mathbf{ks}}\, \hat{c}_\beta^+\, \hat{c}_\alpha \left(\hat{a}_{\mathbf{ks}}\, e^{-i\omega_k t} + \hat{a}_{\mathbf{ks}}^+\, e^{i\omega_k t}\right) e^{i\omega_{\beta\alpha} t} \tag{7.131}$$

with

$$g_{\alpha\beta\mathbf{ks}} = -i \sqrt{\frac{1}{2\epsilon_0 \hbar \omega_k V}}\, \omega_{\beta\alpha}\, \boldsymbol{\varepsilon}_{\mathbf{ks}} \cdot \langle \beta | e\mathbf{r} | \alpha \rangle\,, \tag{7.132}$$

and $\langle \beta | e\mathbf{r} | \alpha \rangle$ the electric dipole moment of the electromagnetic transition between the electronic states $|\beta\rangle$ and $|\alpha\rangle$.

7.6.1 Cavity Quantum Electrodynamics

The development of lasers has produced a new field of research in quantum optics: cavity quantum electrodynamics (CQED). This is the study of the interaction between laser light confined in a reflective cavity and atoms or other particles which are inside the cavity.

The simplest system of CQED is a laser light interacting with a two-level atom. This system can be described by the Hamiltonian (7.119) with Eqs. (7.126), (7.127), (7.128) but considering only one mode of the electromagnetic field and only two modes of the matter field, namely

$$\hat{H} = \hbar\omega\,\hat{a}^+\hat{a} + \epsilon_1\,\hat{c}_1^+\hat{c}_1 + \epsilon_2\,\hat{c}_2^+\hat{c}_2 + \hbar g \left[\hat{c}_2^+\,\hat{c}_1 \left(\hat{a}\,e^{-i(\omega-\omega_{21})t} + \hat{a}^+\,e^{i(\omega+\omega_{21})t}\right)\right.$$
$$\left. + \hat{c}_1^+\,\hat{c}_2 \left(\hat{a}\,e^{-i(\omega+\omega_{21})t} + \hat{a}^+\,e^{i(\omega-\omega_{21})t}\right)\right], \tag{7.133}$$

where for simplicity we set $\omega = \omega_k$, $\omega_{21} = (\epsilon_2 - \epsilon_1)/\hbar$, $\hat{a} = \hat{a}_{\mathbf{ks}}$, $\hat{a}^+ = \hat{a}_{\mathbf{ks}}^+$ and $g = g_{12} = g_{21}$. If the applied electromagnetic radiation is near resonance with an atomic transition, i.e. $\omega \simeq \omega_{21}$ or equivalently $(\omega - \omega_{21}) \simeq 0$, the previous Hamiltonian can be approximated as

$$\hat{H} = \hbar\omega\,\hat{a}^+\hat{a} + \epsilon_1\,\hat{c}_1^+\hat{c}_1 + \epsilon_2\,\hat{c}_2^+\hat{c}_2 + \hbar g \left[\hat{c}_2^+\,\hat{c}_1 \left(\hat{a} + \hat{a}^+\,e^{i(\omega+\omega_{21})t}\right)\right.$$
$$\left. + \hat{c}_1^+\,\hat{c}_2 \left(\hat{a}\,e^{-i(\omega+\omega_{21})t} + \hat{a}^+\right)\right]. \tag{7.134}$$

Moreover, within the so-called rotating-wave approximation, one can neglect the remaining time-dependent terms which oscillate rapidly, obtaining

$$\hat{H} = \hbar\omega\,\hat{a}^+\hat{a} + \epsilon_1\,\hat{c}_1^+\hat{c}_1 + \epsilon_2\,\hat{c}_2^+\hat{c}_2 + \hbar g \left[\hat{c}_2^+\,\hat{c}_1\,\hat{a} + \hat{c}_1^+\,\hat{c}_2\,\hat{a}^+\right]. \tag{7.135}$$

This is the Jaynes-Cummings Hamiltonian, originally proposed in 1963 by Edwin Jaynes and Fred Cummings. This model Hamiltonian can be considered the drosophila of quantum optics: it contains enough physics to describe many phenomena in CQED and atom optics.

Let us analyze some properties of the Jaynes-Cummings Hamiltonian. The photonic Fock state $|n\rangle$ is eigenstate of the bosonic number operator $\hat{a}^+\hat{a}$. The photon can be in any state $|n\rangle$ or or in a superposition of them, namely

$$|\text{photon}\rangle = \sum_{n=0}^{\infty} \alpha_n |n\rangle , \tag{7.136}$$

where α_n are complex coefficients such that $\sum_{n=0}^{\infty} |\alpha_n|^2 = 1$. In addition, the electronic Fock state $|g\rangle = |1_1\rangle$ is eigenstate of the fermionic number operator $\hat{c}_1^+\hat{c}_1$, while the electronic Fock state $|e\rangle = |1_2\rangle$ is eigenstate of the fermionic number operator $\hat{c}_1^+\hat{c}_1$. In this two-level atom the electron can be in the state $|g\rangle$ or in the state $|e\rangle$, or in a superposition of both, namely

$$|\text{electron}\rangle = \alpha|g\rangle + \beta|e\rangle , \tag{7.137}$$

where α and β are two complex coefficients such that $|\alpha|^2 + |\beta|^2 = 1$ and obviously

$$\hat{c}_1^+\hat{c}_1 + \hat{c}_2^+\hat{c}_2 = 1 . \tag{7.138}$$

It is then useful to introduce the following pseudo-spin operators

$$\hat{S}_+ = \hat{c}_2^+\hat{c}_1 , \tag{7.139}$$

$$\hat{S}_- = \hat{c}_1^+\hat{c}_2 , \tag{7.140}$$

$$\hat{S}_z = \frac{1}{2} \left(\hat{c}_2^+\hat{c}_2 - \hat{c}_1^+\hat{c}_1 \right) , \tag{7.141}$$

which have these remarkable properties

$$\hat{S}_+|g\rangle = |e\rangle , \quad \hat{S}_+|e\rangle = 0 , \tag{7.142}$$

$$\hat{S}_-|g\rangle = 0 , \quad \hat{S}_-|e\rangle = |g\rangle , \tag{7.143}$$

$$\hat{S}_z|g\rangle = -\frac{1}{2}|g\rangle , \quad \hat{S}_z|e\rangle = \frac{1}{2}|e\rangle . \tag{7.144}$$

By using these pseudo-spin operators the Jaynes-Cummings Hamiltonian of Eq. (7.135) becomes

$$\hat{H} = \hbar\omega\, \hat{a}^+\hat{a} + \hbar\omega_{21}\hat{S}_z + \hbar g \left(\hat{S}_+\, \hat{a} + \hat{S}_-\, \hat{a}^+ \right) + \bar{\epsilon} , \tag{7.145}$$

where $\bar{\epsilon} = (\epsilon_1 + \epsilon_2)/2$ is a constant energy shift. Due to the deep analogy between two-level electronic system and two-spin states it is quite natural to set

$$| \downarrow \rangle = |g\rangle = |1_1\rangle = \begin{pmatrix} 0 \\ 1 \end{pmatrix} , \tag{7.146}$$

$$| \uparrow \rangle = |e\rangle = |1_2\rangle = \begin{pmatrix} 1 \\ 0 \end{pmatrix} . \tag{7.147}$$

A generic Fock state of the system is then given by

$$|n, \downarrow \rangle = |n\rangle \otimes | \downarrow \rangle \tag{7.148}$$

or

$$|n, \uparrow \rangle = |n\rangle \otimes | \uparrow \rangle . \tag{7.149}$$

We rewrite the Hamiltonian (7.145) neglecting the unrelevant energy shift $\bar{\epsilon}$ as follows

$$\hat{H} = \hat{H}_0 + \hat{H}_I \tag{7.150}$$

where

$$\hat{H}_0 = \hbar\omega \, \hat{a}^+\hat{a} + \hbar\omega_{21}\hat{S}_z \tag{7.151}$$

$$\hat{H}_I = \hbar g \left(\hat{S}_+ \hat{a} + \hat{S}_- \hat{a}^+ \right) . \tag{7.152}$$

Except for the state $|0, \downarrow \rangle$, any other eigenstate of $\hat{H}_0$ belongs to one of the infinite set of two-dimensional subspaces $\{|n - 1, \uparrow \rangle, |n, \downarrow \rangle\}$. Thus, for a fixed and finite value of n, the infinite matrix representing the Jaynes-Cummings Hamiltonian $\hat{H}$ on the basis of eigenstates of $\hat{H}_0$ splis into an infinite set of 2×2 matrices

$$\hat{H}_n = \begin{pmatrix} \hbar\omega \, (n-1) + \hbar\omega_{21}/2 & \hbar g \, \sqrt{n} \\ \hbar g \, \sqrt{n} & \hbar\omega \, n - \hbar\omega_{21}/2 \end{pmatrix} , \tag{7.153}$$

because

$$\hat{S}_+\hat{a}|n - 1, \uparrow \rangle = \sqrt{n-1} \, \hat{S}_+|n - 1, \uparrow \rangle = 0 , \tag{7.154}$$

$$\hat{S}_+\hat{a}|n, \downarrow \rangle = \sqrt{n} \, \hat{S}_+|n - 1, \downarrow \rangle = \sqrt{n} \, |n - 1, \uparrow \rangle , \tag{7.155}$$

$$\hat{S}_-\hat{a}^+|n - 1, \uparrow \rangle = \sqrt{n} \, \hat{S}_-|n, \uparrow \rangle = \sqrt{n} \, |n, \downarrow \rangle , \tag{7.156}$$

$$\hat{S}_-\hat{a}^+|n, \downarrow \rangle = \sqrt{n+1} \, \hat{S}_-|n + 1, \downarrow \rangle = 0 . \tag{7.157}$$

The matrix $\hat{H}_n$ can be easily diagonalized yeldings the following eigenvalues

$$E_n^{(\pm)} = (n - \frac{1}{2})\hbar\omega \pm \frac{1}{2}\Delta \tag{7.158}$$

where
$$\Delta = \sqrt{\delta^2 + 4\hbar^2 g^2 n} \, , \tag{7.159}$$

is the energy splitting with
$$\delta = \hbar(\omega_{21} - \omega) \tag{7.160}$$

the detuning between the atomic (electronic) transition frequency ω_{21} and the photon frequency ω. The corresponding eigenstates are instead given by

$$|u_n^{(+)}\rangle = \cos(\theta)|n, \downarrow\rangle - \sin(\theta)|n-1, \uparrow\rangle \tag{7.161}$$
$$|u_n^{(-)}\rangle = \sin(\theta)|n, \downarrow\rangle + \cos(\theta)|n-1, \uparrow\rangle \, , \tag{7.162}$$

where
$$\theta = \arctan\sqrt{\frac{1 - \delta/\Delta}{1 + \delta/\Delta}} \, . \tag{7.163}$$

The states $|u_n^{(\pm)}\rangle$, known as "dressed states", were introduced by Claude Cohen-Tannoudji and Serge Haroche in 1968–1969. These two scientists got the Nobel Prize in Physics (Cohen-Tannoudji in 1997 and Haroche in 2012) for their studies on the manipulation of atoms with lasers.

7.7 Bosons in a Double-Well Potential

In this last section we consider a quite interesting physical problem: interacting bosons in a quasi-1D double-well potential. The starting point is the quantum-field-theory Hamiltonian

$$\hat{H} = \int d^3\mathbf{r} \, \hat{\psi}^+(\mathbf{r}) \left[-\frac{\hbar^2}{2m}\nabla^2 + U(\mathbf{r}) \right] \hat{\psi}(\mathbf{r})$$
$$+ \frac{1}{2} \int d^3\mathbf{r} \, d^3\mathbf{r}' \, \hat{\psi}^+(\mathbf{r}) \, \hat{\psi}^+(\mathbf{r}') \, V(\mathbf{r} - \mathbf{r}') \, \hat{\psi}(\mathbf{r}') \, \hat{\psi}(\mathbf{r}), \tag{7.164}$$

where the external trapping potential is given by

$$U(\mathbf{r}) = V_{DW}(x) + \frac{1}{2}m\omega_\perp^2(y^2 + z^2) \, , \tag{7.165}$$

that is a generic double-well potential $V_{DW}(x)$ in the x axial direction and a harmonic potential in the transverse (y, z) plane. We assume that the system of bosons, described by the field operator $\hat{\psi}(\mathbf{r})$, is dilute and approximate the inter-particle potential with a contact Fermi pseudo-potential, namely

$$V(\mathbf{r} - \mathbf{r}') = g\,\delta(\mathbf{r} - \mathbf{r}')\,, \tag{7.166}$$

with g the strength of the interaction. If the frequency $\omega_\perp$ is sufficiently large, the system is quasi-1D and the bosonic field operator can be written as

$$\hat{\psi}(\mathbf{r}) = \hat{\phi}(x)\,\frac{e^{-(y^2+z^2)/(2l_\perp^2)}}{\pi^{1/2}l_\perp}\,. \tag{7.167}$$

We are thus supposing that in the transverse (y, z) plane the system is Bose-Einstein condensed into the transverse single-particle ground-state, which is a Gaussian wavefunction of width

$$l_\perp = \sqrt{\frac{\hbar}{m\omega_\perp}}\,, \tag{7.168}$$

that is the characteristic length of the harmonic confinement. Inserting Eq. (7.167) into the Hamiltonian (7.164) and integrating over y and z variabiles we obtain the effective 1D Hamiltonian

$$\hat{H} = \int dx\,\hat{\phi}^+(x)\left[-\frac{\hbar^2}{2m}\frac{d^2}{dx^2} + V_{DW}(x) + \hbar\omega_\perp\right]\hat{\phi}(x)$$
$$+ \frac{g_{1D}}{2}\int dx\,\hat{\phi}^+(x)\,\hat{\phi}^+(x)\,\hat{\phi}(x)\,\hat{\phi}(x)\,, \tag{7.169}$$

where

$$g_{1D} = \frac{g}{2\pi l_\perp^2} \tag{7.170}$$

is the effective 1D interaction strength. Notice that in the one-body part of the Hamiltonian it appears the single-particle ground-state energy $\hbar\omega_\perp$ of the harmonic confinement. We suppose that the barrier of the double-well potential $V_{DW}(x)$, with its maximum located at $x = 0$, is quite high such that there several doublets of quasi-degenerate single-particle energy levels. Moreover, we suppose that only the lowest doublet (i.e. the single-particle ground-state and the single-particle first excited state) is occupied by bosons. Under these assumptions we can write the bosonic field operator as

$$\hat{\phi}(x) = \hat{a}_L\,\phi_L(x) + \hat{a}_R\,\phi_R(x) \tag{7.171}$$

that is the so-called two-mode approximation, where $\phi_L(x)$ and $\phi_R(x)$ are single-particle wavefunctions localized respectively on the left well and on the right well of the double-well potential. These wavefunctions (which can be taken real) are linear combinations of the even (and postitive) wavefunction $\phi_0(x)$ of the ground state and the odd (and positive for $x < 0$) wavefunction $\phi_1(x)$ of the first excited state, namely

$$\phi_L(x) = \frac{1}{\sqrt{2}} \left(\phi_0(x) + \phi_1(x) \right) , \tag{7.172}$$

$$\phi_R(x) = \frac{1}{\sqrt{2}} \left(\phi_0(x) - \phi_1(x) \right) . \tag{7.173}$$

Clearly the operator $\hat{a}_j$ annihilates a boson in the jth site (well) while the operator $\hat{a}_j^+$ creates a boson in the j-th site ($j = L, R$). Inserting the two-mode approximation of the bosonic field operator in the effective 1D Hamiltonian, we get the following two-site Hamiltonian

$$\hat{H} = \epsilon_L \hat{N}_L + \epsilon_R \hat{N}_R - J_{LR} \hat{a}_L^+ \hat{a}_R - J_{RL} \hat{a}_R^+ \hat{a}_L + \frac{U_L}{2} \hat{N}_L(\hat{N}_L - 1) + \frac{U_R}{2} \hat{N}_R(\hat{N}_R - 1) , \tag{7.174}$$

where $\hat{N}_j = \hat{a}_j^+ \hat{a}_j$ is the number operator of the j-th site,

$$\epsilon_j = \int dx \, \phi_j(x) \left[-\frac{\hbar^2}{2m} \frac{d^2}{dx^2} + V_{DW}(x) + \hbar\omega_\perp \right] \phi_j(x) \tag{7.175}$$

is the kinetic plus potential energy on the site j,

$$J_{ij} = \int dx \, \phi_i(x) \left[-\frac{\hbar^2}{2m} \frac{d^2}{dx^2} + V_{DW}(x) + \hbar\omega_\perp \right] \phi_j(x) , \tag{7.176}$$

is the hopping energy (tunneling energy) between the site i and the site j, and

$$U_j = g_{1D} \int dx \, \phi_j(x)^4 \tag{7.177}$$

is the interaction energy on the site j. The Hamiltonian (7.174) is the two-site Bose-Hubbard Hamiltonian, named after John Hubbard introduced a similar model in 1963 to describe fermions (electrons) on a periodic lattice. If the double-well potential $V_{DW}(x)$ is fully symmetric then $\epsilon_L = \epsilon_R = \epsilon$, $J_{LR} = J_{RL} = J$, $U_L = U_R = U$, and the Bose-Hubbard Hamiltonian becomes

$$\hat{H} = \epsilon \left(\hat{N}_L + \hat{N}_R \right) - J \left(\hat{a}_L^+ \hat{a}_R + \hat{a}_R^+ \hat{a}_L \right) + \frac{U}{2} \left[\hat{N}_L(\hat{N}_L - 1) + \hat{N}_R(\hat{N}_R - 1) \right] . \tag{7.178}$$

The Heisenberg equation of motion of the operator $\hat{a}_j$ is given by

$$i\hbar \frac{d}{dt} \hat{a}_j = [\hat{a}_j, \hat{H}] = \frac{\partial \hat{H}}{\partial \hat{a}_j^+} , \tag{7.179}$$

from which we obtain

$$i\hbar \frac{d}{dt} \hat{a}_j = \epsilon \hat{a}_j - J \hat{a}_i + U(\hat{N}_j - \frac{1}{2}) \hat{a}_j , \tag{7.180}$$

where $j = L, R$ and $i = R, L$. By averaging this equation with the coherent state $|\alpha_L \alpha_R\rangle = |\alpha_L\rangle \otimes |\alpha_R\rangle$, such that $\hat{a}_j(t)|\alpha_j\rangle = \alpha_j(t)|\alpha_j\rangle$, we find

$$i\hbar\frac{d}{dt}\alpha_j = \epsilon\,\alpha_j - J\alpha_i + U(|\alpha_j|^2 - \frac{1}{2})\alpha_j , \qquad (7.181)$$

where

$$\alpha_j(t) = \bar{N}_j(t)\,e^{i\theta_j(t)} , \qquad (7.182)$$

with $\bar{N}_j(t)$ the average number of bosons in the site j at time t and $\theta_j(t)$ the corresponding phase angle at the same time t. Working with a fixed number of bosons, i.e.

$$N = \bar{N}_L(t) + \bar{N}_R(t) , \qquad (7.183)$$

and introducing population imbalance

$$z(t) = \frac{\bar{N}_L(t) - \bar{N}_R(t)}{N} \qquad (7.184)$$

and phase difference

$$\theta(t) = \theta_R(t) - \theta_L(t) , \qquad (7.185)$$

the time-dependent equations for $\alpha_L(t)$ and $\alpha_R(t)$ can be re-written as follows

$$\frac{dz}{dt} = -\frac{2J}{\hbar}\sqrt{1 - z^2}\,\sin(\theta) , \qquad (7.186)$$

$$\frac{d\theta}{dt} = \frac{2J}{\hbar}\frac{z}{\sqrt{1 - z^2}}\cos(\theta) + \frac{U}{\hbar}z . \qquad (7.187)$$

These are the so-called Josephson equations of the Bose-Einstein condensate. Indeed, under the condition of small population imbalance ($|z| \ll 1$), small on-site interaction energy ($|U|z/\hbar \ll 1$) and small phase difference ($\theta \ll 1$) one finds

$$\frac{dz}{dt} = -\frac{2J}{\hbar}\sin(\theta) , \qquad (7.188)$$

$$\frac{d\theta}{dt} = \frac{2J}{\hbar}z , \qquad (7.189)$$

which are the equations introduced in 1962 by Brian Josephson to describe the superconducting electric current (made of quasi-bosonic Cooper pairs of electrons) between two superconductors separated by a thin insulating barrier (Nobel Prize in 1973). Remarkably, as shown in 2007 at Heidelberg by the experimental group of Markus Oberthaler, Eqs. (7.186) and (7.187) describe accurately also the dynamics of a Bose-Einstein condensate made of alkali-metal atoms confined in the quasi-1D double-well potential by counter-propagating laser beams.

7.7.1 Analytical Results with $N = 1$ and $N = 2$

In this subsection we discuss the two-site Bose-Hubbard Hamiltonian

$$\hat{H} = -J \left(\hat{a}_L^+ \hat{a}_R + \hat{a}_R^+ \hat{a}_L \right) + \frac{U}{2} \left[\hat{N}_L(\hat{N}_L - 1) + \frac{U_R}{2} \hat{N}_R(\hat{N}_R - 1) \right] , \quad (7.190)$$

in the case of $N = 1$ and $N = 2$ bosons. Note that, without loss of generality, we have removed the constant energy $\epsilon \hat{N}$.

The ground-state $|GS\rangle$ of the system with N bosons can be written as

$$|GS\rangle = \sum_{i=0}^{N} c_i \, |i, N - i\rangle , \quad (7.191)$$

where $|i, N - i\rangle = |i\rangle \otimes |N - i\rangle$ is the state with i bosons in the left well and $N - i$ bosons in the right well and c_i is its probability amplitude such that

$$\sum_{i=0}^{N} |c_i|^2 = 1 . \quad (7.192)$$

In the case of $N = 1$ bosons the ground-state is simply

$$|GS\rangle = c_0|0, 1\rangle + c_1|1, 0\rangle , \quad (7.193)$$

where

$$|c_0|^2 + |c_1|^2 = 1 . \quad (7.194)$$

Moreover, the on-site energy is identically zero and consequently it is immediate to find

$$c_0 = c_1 = \frac{1}{\sqrt{2}} \quad (7.195)$$

for any value of J and U.

In the case of $N = 2$ bosons the ground-state is instead given by

$$|GS\rangle = c_0|0, 2\rangle + c_1|1, 1\rangle + c_2|2, 0\rangle , \quad (7.196)$$

where

$$|c_0|^2 + |c_1|^2 + |c_2|^2 = 1 . \quad (7.197)$$

By direct diagonalization of the corresponding Hamiltonian matrix, one finds

$$c_0 = c_2 = \frac{2}{\sqrt{16 + \zeta^2 + \zeta\sqrt{\zeta^2 + 16}}} \tag{7.198}$$

and

$$c_1 = c_0 \frac{\zeta + \sqrt{\zeta^2 + 16}}{2\sqrt{2}}, \tag{7.199}$$

where

$$\zeta = \frac{U}{J} \tag{7.200}$$

is the effective dimensionless interaction strength which controls the nature of the ground state $|GS\rangle$. Thus, the ground-state is strongly dependent on the ratio between the on-site energy U and the tunneling (hopping) energy J. In particular, one deduces that

$$|GS\rangle = \begin{cases} |1, 1\rangle & \text{for } \zeta \to +\infty \\ \frac{1}{2}\left(|0, 2\rangle + \sqrt{2}|1, 1\rangle + |2, 0\rangle\right) & \text{for } \zeta \to 0 \\ \frac{1}{\sqrt{2}}\left(|0, 2\rangle + |2, 0\rangle\right) & \text{for } \zeta \to -\infty \end{cases}. \tag{7.201}$$

7.8 Solved Problems

Problem 7.1
Consider the operator $\hat{N} = \hat{f}^+\hat{f}$, where $\hat{f}$ and $\hat{f}^+$ satisfy the anti-commutation rule $\hat{f}\hat{f}^+ + \hat{f}^+\hat{f} = 1$. Show that if $|n\rangle$ is an eigenstate of $\hat{N}$ with eigenvalue n then $\hat{f}^+|n\rangle$ is eigenstates of $\hat{N}$ with eigenvalue $1 - n$.

Solution
We have

$$\hat{N}\hat{f}|n\rangle = (\hat{f}^+\hat{f})\hat{f}|n\rangle.$$

The anti-commutation relation between $\hat{f}$ and $\hat{f}^+$ can be written as

$$\hat{f}\hat{f}^+ = 1 - \hat{f}^+\hat{f}.$$

This implies that

$$\hat{N}\hat{f}^+|n\rangle = (\hat{f}^+\hat{f})\hat{f}^+|n\rangle = \hat{f}^+(\hat{f}\hat{f}^+)|n\rangle = \hat{f}^+(1 - \hat{f}^+\hat{f})|n\rangle = \hat{f}^+(1 - \hat{N})|n\rangle = (1-n)\hat{f}^+|n\rangle.$$

Notice that in Sect. 7.2 we have shown that the operator $\hat{N}$ of fermions has only eigenvalues 0 and 1. This means that $\hat{f}^+|0\rangle = |1\rangle$ and $\hat{f}^+|1\rangle = 0$.

Problem 7.2

Show that the following equation of motion

$$i\hbar\frac{\partial}{\partial t}\psi(\mathbf{r}, t) = \left[-\frac{\hbar^2}{2m}\nabla^2 + U(\mathbf{r})\right]\psi(\mathbf{r}, t) + g\,|\psi(\mathbf{r}, t)|^2\psi(\mathbf{r}, t)$$

of the classical Schrödinger field $\psi(\mathbf{r}, t)$, that is the time-dependent Gross-Pitaevskii equation, can be deduced by extremizing the action

$$S = \int dt d^3\mathbf{r}\, \mathcal{L}\,,$$

where

$$\mathcal{L} = \psi^*(\mathbf{r}, t)\left[i\hbar\frac{\partial}{\partial t} + \frac{\hbar^2}{2m}\nabla^2 - U(\mathbf{r})\right]\psi(\mathbf{r}, t) - \frac{1}{2}g\,|\psi(\mathbf{r}, t)|^4$$

is the Lagrangian density of the system.

Solution

First of all, we observe that

$$\nabla \cdot \left(\psi^*\nabla\psi\right) = \psi^*\nabla^2\psi + \nabla\psi^* \cdot \nabla\psi\,.$$

We can then write

$$\int_V d^3\mathbf{r}\,\psi^*\nabla^2\psi = \int_V d^3\mathbf{r}\,\nabla \cdot \left(\psi^*\nabla\psi\right) - \int_V d^3\mathbf{r}\,\nabla\psi^* \cdot \nabla\psi$$

$$= \int_S d^2\mathbf{r}\,\left(\psi^*\nabla\psi\right) \cdot \mathbf{n} - \int_V d^3\mathbf{r}\,\nabla\psi^* \cdot \nabla\psi\,,$$

where S is the surface of the domain of volume V. In the limit of a very large V we can suppose that the Schrödinger field and its derivatives are zero on the surface S, and consequently

$$\int_V d^3\mathbf{r}\,\psi^*\nabla^2\psi = -\int_V d^3\mathbf{r}\,\nabla\psi^* \cdot \nabla\psi\,.$$

This means that the Lagrangian density can be rewritten as

$$\mathcal{L} = i\hbar\psi^*\frac{\partial\psi}{\partial t} - \frac{\hbar^2}{2m}|\nabla\psi|^2 - U(\mathbf{r})|\psi|^2 - \frac{1}{2}g|\psi|^4\,,$$

where $|\nabla\psi|^2 = \nabla\psi^* \cdot \nabla\psi$. In this way the Lagrangian density is a function of the Schrödinger field and only its first derivatives, namely

$$\mathcal{L} = \mathcal{L}(\psi, \psi^*, \frac{\partial\psi}{\partial t}, \nabla\psi, \nabla\psi)\,.$$

The Schrödinger field $\psi(\mathbf{r}, t) = \psi_R(\mathbf{r}, t) + i\psi_I(\mathbf{r}, t)$ is complex, depending on the two real fields $\psi_R(\mathbf{r}, t)$ and $\psi_I(\mathbf{r}, t)$ which can be varied independently, but in this way the complex conjugate field $\psi^*(\mathbf{r}, t) = \psi_R(\mathbf{r}, t) - i\psi_I(\mathbf{r}, t)$ is fully determined. Alternatively, one can consider $\psi(\mathbf{r}, t)$ and $\psi^*(\mathbf{r}, t)$ as independent fields: this is our choice and the action S is then a functional of $\psi(\mathbf{r}, t)$ and $\psi(\mathbf{r}, t)$. The first variation of the action S gives

$$\delta S = \int dt d^3\mathbf{r} \left(\frac{\delta S}{\delta \psi} \delta\psi + \frac{\delta S}{\delta \psi} \delta\psi^* \right) ,$$

where

$$\frac{\delta S}{\delta \psi} = \frac{\partial \mathcal{L}}{\partial \psi} - \frac{\partial}{\partial t} \frac{\partial \mathcal{L}}{\partial \frac{\partial \psi}{\partial t}} - \mathbf{\nabla} \cdot \frac{\partial \mathcal{L}}{\partial \mathbf{\nabla} \psi} ,$$

$$\frac{\delta S}{\delta \psi^*} = \frac{\partial \mathcal{L}}{\partial \psi^*} - \frac{\partial}{\partial t} \frac{\partial \mathcal{L}}{\partial \frac{\partial \psi^*}{\partial t}} - \mathbf{\nabla} \cdot \frac{\partial \mathcal{L}}{\partial \mathbf{\nabla} \psi^*} .$$

Imposing that

$$\delta S = 0$$

we obtain two Euler-Lagrange equations

$$\frac{\partial \mathcal{L}}{\partial \psi} - \frac{\partial}{\partial t} \frac{\partial \mathcal{L}}{\partial \frac{\partial \psi}{\partial t}} - \mathbf{\nabla} \cdot \frac{\partial \mathcal{L}}{\partial \mathbf{\nabla} \psi} = 0 ,$$

$$\frac{\partial \mathcal{L}}{\partial \psi^*} - \frac{\partial}{\partial t} \frac{\partial \mathcal{L}}{\partial \frac{\partial \psi^*}{\partial t}} - \mathbf{\nabla} \cdot \frac{\partial \mathcal{L}}{\partial \mathbf{\nabla} \psi^*} = 0 .$$

By using our Lagrangian density the first Euler-Lagrange equation gives

$$i\hbar \frac{\partial}{\partial t} \psi = \left[-\frac{\hbar^2}{2m} \nabla^2 + U(\mathbf{r}) \right] \psi + g |\psi|^2 \psi ,$$

while the second Euler-Lagrange equation gives

$$-i\hbar \frac{\partial}{\partial t} \psi^* = \left[-\frac{\hbar^2}{2m} \nabla^2 + U(\mathbf{r}) \right] \psi^* + g |\psi|^2 \psi^* ,$$

that is the complex conjugate of the previous one.

To conclude, we observe that if the Lagrangian density depends also on higher spatial derivatives the Euler-Lagrange equations are modified accordingly. For instance, in the presence of terms like $\nabla^2 \psi$ or $\nabla^2 \psi^*$ one gets

$$\frac{\partial \mathcal{L}}{\partial \psi} - \frac{\partial}{\partial t} \frac{\partial \mathcal{L}}{\partial \frac{\partial \psi}{\partial t}} - \nabla \cdot \frac{\partial \mathcal{L}}{\partial \nabla \psi} + \nabla^2 \frac{\partial \mathcal{L}}{\partial \nabla^2 \psi} = 0 \,,$$

$$\frac{\partial \mathcal{L}}{\partial \psi^*} - \frac{\partial}{\partial t} \frac{\partial \mathcal{L}}{\partial \frac{\partial \psi^*}{\partial t}} - \nabla \cdot \frac{\partial \mathcal{L}}{\partial \nabla \psi^*} + \nabla^2 \frac{\partial \mathcal{L}}{\partial \nabla^2 \psi^*} = 0 \,.$$

Problem 7.3

Derive the Gross-Pitaevskii equation

$$i\hbar \frac{\partial}{\partial t} \psi_0(\mathbf{r}, t) = \left[-\frac{\hbar^2}{2m} \nabla^2 + U(\mathbf{r}) \right] \psi_0(\mathbf{r}, t) + g\bar{N} \, |\psi_0(\mathbf{r}, t)|^2 \psi_0(\mathbf{r}, t)$$

of the classical field $\psi_0(\mathbf{r}, t)$ from the Heisenberg equation of motion

$$i\hbar \frac{\partial}{\partial t} \hat{\psi}(\mathbf{r}, t) = \left[-\frac{\hbar^2}{2m} \nabla^2 + U(\mathbf{r}) \right] \hat{\psi}(\mathbf{r}, t) + g \, \hat{\psi}^+(\mathbf{r}, t)\hat{\psi}(\mathbf{r}, t)\hat{\psi}(\mathbf{r}, t)$$

of the bosonic quantum field operator $\hat{\psi}(\mathbf{r}, t)$.
Suggestion: expand the quantum field operator as

$$\hat{\psi}(\mathbf{r}, t) = \sum_\alpha \hat{c}_\alpha \, \phi_\alpha(\mathbf{r}, t)$$

and use the coherent state $|c_0\rangle$ of $\hat{c}_0$.

Solution

First of all we notice that

$$\begin{aligned}
\langle c_0 | i\hbar \frac{\partial}{\partial t} \hat{\psi}(\mathbf{r}, t) | c_0 \rangle &= i\hbar \frac{\partial}{\partial t} \langle c_0 | \hat{\psi}(\mathbf{r}, t) | c_0 \rangle \\
&= i\hbar \frac{\partial}{\partial t} \langle c_0 | \sum_\alpha \hat{c}_\alpha \, \phi_\alpha(\mathbf{r}, t) | c_0 \rangle \\
&= i\hbar \frac{\partial}{\partial t} \sum_\alpha \phi_\alpha(\mathbf{r}, t) \langle c_0 | \hat{c}_\alpha | c_0 \rangle \\
&= i\hbar \frac{\partial}{\partial t} \sum_\alpha \phi_\alpha(\mathbf{r}, t) c_0 \, \delta_{\alpha, c_0} \\
&= c_0 \, i\hbar \frac{\partial}{\partial t} \phi_0(\mathbf{r}, t) \,.
\end{aligned}$$

Similarly

$$\langle c_0 | \left[-\frac{\hbar^2}{2m} \nabla^2 + U(\mathbf{r}) \right] \hat{\psi}(\mathbf{r}, t) | c_0 \rangle = \left[-\frac{\hbar^2}{2m} \nabla^2 + U(\mathbf{r}) \right] \langle c_0 | \hat{\psi}(\mathbf{r}, t) | c_0 \rangle$$

$$= \left[-\frac{\hbar^2}{2m} \nabla^2 + U(\mathbf{r}) \right] \langle c_0 | \sum_\alpha \hat{c}_\alpha \, \phi_\alpha(\mathbf{r}, t) | c_0 \rangle$$

$$= \left[-\frac{\hbar^2}{2m} \nabla^2 + U(\mathbf{r}) \right] \sum_\alpha \phi_\alpha(\mathbf{r}, t) \langle c_0 | \hat{c}_\alpha | c_0 \rangle$$

$$= \left[-\frac{\hbar^2}{2m} \nabla^2 + U(\mathbf{r}) \right] \sum_\alpha \phi_\alpha(\mathbf{r}, t) c_0 \, \delta_{\alpha,0}$$

$$= c_0 \left[-\frac{\hbar^2}{2m} \nabla^2 + U(\mathbf{r}) \right] \phi_0(\mathbf{r}, t) \ .$$

Moreover

$$\langle c_0 | \hat{\psi}^+(\mathbf{r}, t) \hat{\psi}(\mathbf{r}, t) \hat{\psi}(\mathbf{r}, t) | c_0 \rangle = \langle c_0 | \sum_\alpha \hat{c}_\alpha^+ \, \phi_\alpha^*(\mathbf{r}, t) \sum_\beta \hat{c}_\beta \, \phi_\beta(\mathbf{r}, t) \sum_\gamma \hat{c}_\gamma \, \phi_\gamma(\mathbf{r}, t) | c_0 \rangle$$

$$= \sum_{\alpha\beta\gamma} \phi_\alpha^*(\mathbf{r}, t) \phi_\beta(\mathbf{r}, t) \phi_\gamma(\mathbf{r}, t) \langle c_0 | \hat{c}_\alpha^+ \hat{c}_\beta \hat{c}_\gamma | c_0 \rangle$$

$$= \sum_{\alpha\beta\gamma} \phi_\alpha^*(\mathbf{r}, t) \phi_\beta(\mathbf{r}, t) \phi_\gamma(\mathbf{r}, t) |c_0|^2 c_0 \delta_{\alpha,0} \delta_{\beta,0} \delta_{\gamma,0}$$

$$= |c_0|^2 c_0 |\phi_0(\mathbf{r}, t)|^2 \phi_0(\mathbf{r}, t) \ .$$

Finally, after setting $\bar{N} = |c_0|^2$, we conclude that

$$\langle c_0 | i\hbar \frac{\partial}{\partial t} \hat{\psi}(\mathbf{r}, t) | c_0 \rangle = \langle c_0 | \left[-\frac{\hbar^2}{2m} \nabla^2 + U(\mathbf{r}) \right] \hat{\psi}(\mathbf{r}, t) + g \, \hat{\psi}^+(\mathbf{r}, t) \hat{\psi}(\mathbf{r}, t) \hat{\psi}(\mathbf{r}, t) | c_0 \rangle$$

is exactly

$$i\hbar \frac{\partial}{\partial t} \psi_0(\mathbf{r}, t) = \left[-\frac{\hbar^2}{2m} \nabla^2 + U(\mathbf{r}) \right] \psi_0(\mathbf{r}, t) + g\bar{N} \, |\psi_0(\mathbf{r}, t)|^2 \psi_0(\mathbf{r}, t) \ .$$

Problem 7.4
Calculate the Brezin integral

$$\int d\bar{c} \, dc \, e^{-\bar{c}Ac} \ ,$$

where c and $\bar{c}$ are Grassmann variables, and A is a complex number.

Solution
First of all we notice that, due to the fact that we are working with Grasmann numbers, one finds

$$e^{-\bar{c}Ac} = 1 - \bar{c}Ac \; ,$$

because all the other terms of the Taylor-MacLaurin expansion are identically zero. Consequently

$$\int d\bar{c}\,dc\; e^{-\bar{c}Ac} = \int d\bar{c}\,dc\; (1 - \bar{c}Ac) \; .$$

Finally, on the basis of the definition of the Brezin integral where the integration is defined as equivalent to differentiation, we get

$$\int d\bar{c}\,dc\; (1 - \bar{c}Ac) = \frac{\partial^2}{\partial\bar{c}\partial c} (1 - \bar{c}Ac) = A \; .$$

Further Reading

For the second quantization of particles and non-relativistic quantum field theory:
N. Nagaosa, Quantum Field Theory in Condensed Matter Physics, Chap. 1, Sects. 1.1 and 1.2 (Springer, Berlin, 1999).
A.L. Fetter and J.D. Walecka, Quantum Theory of Many-Particle Systems, Chap. 1, Sects. 1.1 and 1.2 (Dover Publications, New York, 2003).
H.T.C. Stoof, K.B. Gubbels, and D.B.M. Dickerscheid, Ultracold Quantum Fields, Chap. 6, Sects. 6.1, 6.2, 6.3, and 6.4 (Springer, Berlin, 2009).
A. Atland and B. Simons, Condensed Matter Field Theory, Chap. 2, Sect. 2.1 (Cambridge Univ. Press, Cambridge, 2006).
For the quantum properties of bosons in a double-well potential:
G. Mazzarella, L. Salasnich, A. Parola, and F. Toigo, Coherence and entanglement in the ground-state of a bosonic Josephson junction:from macroscopic Schrdinger cats to separable Fock states, Phys. Rev. A **83**, 053607 (2011).

Chapter 8
Functional Integration for the Bosonic Field

In this chapter we consider the quantum statistical mechanics of identical bosons by introducing the formalism of functional integration for the bosonic field. We then derive the equation of state of non-interacting and weakly-interacting bosons taking into account beyond-mean-field regularized Gaussian fluctuations. We also analyze various properties of the mean-field time-dependent Gross-Pitaevskii equation. In particular we show that this equation is equivalent to the zero-temperature equations of superfluid hydrodynamics proposed in 1949 by Landau. Finally, we discuss topological and solitonic solutions of the Gross-Pitaevskii equation.

8.1 Partition Function of Interacting Bosons

For a system of non-relativistic interacting bosons the partition function is given by

$$\mathcal{Z} = Tr[e^{-\beta(\hat{H} - \mu \hat{N})}] \,, \tag{8.1}$$

where the second-quantized Hamiltonian is

$$
\begin{aligned}
\hat{H} = \int d^3\mathbf{r}\, \hat{\psi}^+(\mathbf{r})\left(-\frac{\hbar^2}{2m}\nabla^2 + U(\mathbf{r})\right)\hat{\psi}(\mathbf{r}) \\
+ \frac{1}{2}\int d^3\mathbf{r}\, d^3\mathbf{r}'\, \hat{\psi}^+(\mathbf{r})\, \hat{\psi}^+(\mathbf{r}')\, V(\mathbf{r}, \mathbf{r}')\, \hat{\psi}(\mathbf{r}')\, \hat{\psi}(\mathbf{r}) \,,
\end{aligned}
\tag{8.2}
$$

with $U(\mathbf{r})$ the external potential acting on spinless bosons described by the field operator $\hat{\psi}(\mathbf{r})$, $V(\mathbf{r}, \mathbf{r}')$ the inter-particle interaction, and the number operator reads

$$\hat{N} = \int d^3\mathbf{r}\, \hat{\psi}^+(\mathbf{r})\hat{\psi}(\mathbf{r}) \,. \tag{8.3}$$

© Springer International Publishing AG 2017

L. Salasnich, *Quantum Physics of Light and Matter*, UNITEXT for Physics,
DOI 10.1007/978-3-319-52998-1_8

In Eq. (8.1), as usual, $\beta = 1/(k_B T)$ with k_B the Boltzmann constant and T the temperature, while μ is the chemical potential.

In the previous chapter we have seen that, by expanding the field operator $\hat{\psi}(\mathbf{r})$ as

$$\hat{\psi}(\mathbf{r}) = \sum_\alpha \hat{c}_\alpha \, \phi_\alpha(\mathbf{r}) \,, \tag{8.4}$$

where $\phi_\alpha(\mathbf{r})$ are the orthonormal eigenfunctions of the non-interacting problem with single-particle energies ϵ_α, one finds

$$\hat{H} = \sum_\alpha \epsilon_\alpha \, \hat{c}_\alpha^+ \hat{c}_\alpha + \sum_{\alpha\beta\gamma\delta} V_{\alpha\beta\gamma\delta} \, \hat{c}_\alpha^+ \hat{c}_\beta^+ \hat{c}_\delta \hat{c}_\gamma \,, \tag{8.5}$$

where

$$V_{\alpha\beta\delta\gamma} = \int d^3\mathbf{r} \, d^3\mathbf{r}' \, \phi_\alpha^*(\mathbf{r}) \, \phi_\beta^*(\mathbf{r}') \, V(\mathbf{r}, \mathbf{r}') \, \phi_\delta(\mathbf{r}') \, \phi_\gamma(\mathbf{r}) \,, \tag{8.6}$$

and also

$$\hat{N} = \sum_\alpha \hat{c}_\alpha^+ \hat{c}_\alpha \,. \tag{8.7}$$

Moreover, we have seen that the generalized coherent state

$$|\psi\rangle = \prod_\alpha |c_\alpha\rangle = e^{-\sum_\alpha |c_\alpha|^2/2} \, e^{\sum_\alpha c_\alpha \hat{c}_\alpha^+} |0\rangle \,, \tag{8.8}$$

is such that

$$\hat{\psi}(\mathbf{r}) |\psi\rangle = \psi(\mathbf{r}) |\psi\rangle \,, \tag{8.9}$$

where

$$\psi(\mathbf{r}) = \sum_\alpha c_\alpha \, \phi_\alpha(\mathbf{r}) \tag{8.10}$$

is the "classical" Schrödinger field with c_α the complex eigenvalues of $\hat{c}_\alpha$. Clearly, it follows that

$$\hat{c}_\alpha |\psi\rangle = c_\alpha |\psi\rangle \quad \text{for any } \alpha \,. \tag{8.11}$$

Note that $|\psi\rangle$ is the tensor product of all single-mode coherent states $|c_\alpha\rangle$. Due to their normalization we get

$$\langle \psi | \psi \rangle = 1 \,. \tag{8.12}$$

The overlap between two generalized coherent states $|\psi\rangle$ and $|\tilde{\psi}\rangle$ is instead given by

$$\langle \psi | \tilde{\psi} \rangle = e^{-\frac{1}{2} \sum_\alpha [c_\alpha^*(c_\alpha - \tilde{c}_\alpha) - (c_\alpha^* - \tilde{c}_\alpha^*)\tilde{c}_\alpha]} \,. \tag{8.13}$$

The single-mode coherent state $|c_\alpha\rangle$ satisfies the remarkable completeness relation

$$\int \frac{dc_\alpha^* dc_\alpha}{2\pi i} |c_\alpha\rangle\langle c_\alpha| = 1 \ . \tag{8.14}$$

It is possible to prove that the set of single-mode coherent states $\{|c_\alpha\rangle\}$ obtained by varying α is over-complete and the generalized coherent state $|\psi\rangle$ satisfies the completeness relation

$$\int d[\psi^*, \psi] \, |\psi\rangle\langle\psi| = 1 \ , \tag{8.15}$$

where the integration measure is defined by

$$\int d[\psi^*, \psi] = \int \prod_\alpha \frac{dc_\alpha^* dc_\alpha}{2\pi i} \tag{8.16}$$

with $\psi^*(\mathbf{r})$ the complex conjugate of $\psi(\mathbf{r})$ such that

$$\langle\psi| \hat{\psi}^+(\mathbf{r}) = \psi^*(\mathbf{r}) \langle\psi| \ . \tag{8.17}$$

Notice that c_α and $\psi(\mathbf{r})$ represent the same complex field in reciprocal spaces, and the integration measure can also be written as

$$\int d[\psi^*, \psi] = \int \prod_\mathbf{r} \frac{d\psi^*(\mathbf{r}) d\psi(\mathbf{r})}{2\pi i} \ . \tag{8.18}$$

Moreover, Eq. (8.13) can also be written as

$$\langle\psi|\tilde{\psi}\rangle = e^{-\frac{1}{2} \int d^3\mathbf{r} [\psi^*(\mathbf{r})(\psi(\mathbf{r})-\tilde{\psi}(\mathbf{r})) - (\psi^*(\mathbf{r})-\tilde{\psi}^*(\mathbf{r}))\tilde{\psi}(\mathbf{r})]} \ . \tag{8.19}$$

The key point is that the partition function $\mathcal{Z}$ can be expressed as a functional integral taking into account that the trace of an operator can be performed by using any set of basis states, and in particular the basis of generalized coherent states. Thus, we write

$$\mathcal{Z} = Tr[e^{-\beta(\hat{H}-\mu\hat{N})}] = \int d[\psi^*, \psi] \, \langle\psi|e^{-\beta(\hat{H}-\mu\hat{N})}|\psi\rangle \ , \tag{8.20}$$

where $|\psi\rangle$ is the generalized coherent state.

8.1.1 Semiclassical Approximation and Imaginary Time

The main problem in the functional representation of the partition function $\mathcal{Z}$ is the calculation of

$$\langle\psi|e^{-\beta(\hat{H}-\mu\hat{N})}|\psi\rangle \ . \tag{8.21}$$

The simple semiclassical approximation

$$\langle \psi | e^{-\beta(\hat{H}-\mu\hat{N})} | \psi \rangle \simeq e^{-\beta\langle\psi|(\hat{H}-\mu\hat{N})|\psi\rangle} \tag{8.22}$$

gives

$$\mathcal{Z} = \int d[\psi^*, \psi] e^{-\beta(E[\psi^*,\psi]-\mu N[\psi^*,\psi])} , \tag{8.23}$$

where

$$E[\psi^*, \psi] = \int d^3\mathbf{r}\, \psi^*(\mathbf{r}) \left(-\frac{\hbar^2}{2m}\nabla^2 + U(\mathbf{r}) \right) \psi(\mathbf{r})$$
$$+ \frac{1}{2} \int d^3\mathbf{r}\, d^3\mathbf{r}'\, |\psi(\mathbf{r})|^2 \, V(\mathbf{r}, \mathbf{r}') \, |\psi(\mathbf{r}')|^2 \tag{8.24}$$

and

$$N[\psi^*, \psi] = \int d^3\mathbf{r}\, \psi^*(\mathbf{r})\psi(\mathbf{r}) . \tag{8.25}$$

The partition function given by Eq. (8.23) is called semiclassical because it is usually reliable only at high temperatures, where quantum statistics plays a marginal role.

To better treat the expectation value of Eq. (8.21) we first write $|\psi_0\rangle$ instead of $|\psi\rangle$ and also (with $M \to \infty$)

$$\Delta\tau = \frac{\hbar\beta}{M} . \tag{8.26}$$

As we shall see immediately, this number $\Delta\tau$ plays the role of the step interval of an imaginary time τ. In fact, we obtain

$$\langle\psi_0| e^{-\beta(\hat{H}-\mu\hat{N})}|\psi_0\rangle = \langle\psi_0| e^{-M\Delta\tau(\hat{H}-\mu\hat{N})/\hbar}|\psi_0\rangle$$
$$= \langle\psi_0| \left(e^{-\Delta\tau(\hat{H}-\mu\hat{N})/\hbar} \right)^M |\psi_0\rangle \tag{8.27}$$
$$= \langle\psi_0| e^{-\Delta\tau(\hat{H}-\mu\hat{N})/\hbar} \cdots e^{-\Delta\tau(\hat{H}-\mu\hat{N})/\hbar}|\psi_0\rangle .$$

Then we insert $M-1$ completeness relations for the general coherent state $|\psi_j\rangle$ ($j = 1, ..., M-1$)

$$\int d[\psi_j^*, \psi_j] |\psi_j\rangle\langle\psi_j| = 1 \tag{8.28}$$

to get

$$\langle\psi_0| e^{-\beta(\hat{H}-\mu\hat{N})}|\psi_0\rangle = \int \left(\prod_{j=1}^{M-1} d[\psi_j^*, \psi_j] \right) \prod_{j=1}^{M} \langle\psi_j| e^{-\Delta\tau(\hat{H}-\mu\hat{N})/\hbar}|\psi_{j-1}\rangle \tag{8.29}$$

imposing that $|\psi_0\rangle = |\psi_M\rangle$. In the limit $M \to \infty$ one has $\Delta\tau \to 0$ and also

$$\langle \psi_j | e^{-\Delta\tau(\hat{H}-\mu\hat{N})/\hbar} | \psi_{j-1} \rangle = \langle \psi_j | \psi_{j-1} \rangle \, e^{-\Delta\tau(E[\psi_j^*,\psi_j]-\mu N[\psi_j^*,\psi_j])/\hbar} \ . \tag{8.30}$$

Moreover, from Eq. (8.19) the overlap between two general coherent states reads

$$\langle \psi_j | \psi_{j-1} \rangle = e^{-\frac{\Delta\tau}{2} \int d^3\mathbf{r} \left(\psi_j^*(\mathbf{r}) \frac{[\psi_j(\mathbf{r})-\psi_{j-1}(\mathbf{r})]}{\Delta\tau} - \psi_j(\mathbf{r}) \frac{[\psi_j^*(\mathbf{r})-\psi_{j-1}^*(\mathbf{r})]}{\Delta\tau} \right)} \ . \tag{8.31}$$

Finally, setting $\tau = j\Delta\tau$, $\psi(\mathbf{r},\tau) = \psi_j(\mathbf{r})$ and

$$\frac{\partial}{\partial\tau}\psi(\mathbf{r},\tau) = \frac{[\psi_j(\mathbf{r}) - \psi_{j-1}(\mathbf{r})]}{\Delta\tau} \ , \tag{8.32}$$

we obtain the partition function

$$\mathcal{Z} = \int \mathcal{D}[\psi,\psi^*] \, \exp\left\{ -\frac{S[\psi,\psi^*]}{\hbar} \right\} \ , \tag{8.33}$$

where

$$S[\psi,\psi^*] = \int_0^{\hbar\beta} d\tau \int d^3\mathbf{r} \, \mathcal{L}(\psi^*,\psi) \tag{8.34}$$

is the Euclidean (i.e. with imaginary time) action of the Euclidean Lagrangian density

$$\begin{aligned}
\mathcal{L}(\psi^*,\psi) = \ &\frac{\hbar}{2}\left(\psi^*(\mathbf{r},\tau)\frac{\partial}{\partial\tau}\psi(\mathbf{r},\tau) - \psi(\mathbf{r},\tau)\frac{\partial}{\partial\tau}\psi^*(\mathbf{r},\tau) \right) \\
&+ \psi^*(\mathbf{r},\tau)\left(-\frac{\hbar^2}{2m}\nabla^2 + U(\mathbf{r}) - \mu \right)\psi(\mathbf{r},\tau) \\
&+ \frac{1}{2}\int d^3\mathbf{r}' \, |\psi(\mathbf{r}',\tau)|^2 \, V(\mathbf{r},\mathbf{r}') \, |\psi(\mathbf{r},\tau)|^2
\end{aligned} \tag{8.35}$$

with time-periodic boundary conditions $\psi(\mathbf{r},0) = \psi(\mathbf{r},\hbar\beta)$ and

$$\int \mathcal{D}[\psi,\psi^*] = \int \prod_{(\mathbf{r},\tau)} \frac{d\psi^*(\mathbf{r},\tau)\,d\psi(\mathbf{r},\tau)}{2\pi i} \ . \tag{8.36}$$

We stress again that a semiclassical treatment of the expectation value (8.21) gives a semiclassical partition function (8.23), which is written as a functional integration of the time-independent energy functional. Instead, the exact treatment of Eq. (8.21) produces the exact partition function (8.33), which is expressed as a functional integration of the imaginary-time action functional.

8.2 Broken Symmetry and Order Parameter

We consider a D-dimensional ($D = 1, 2, 3$) Bose gas either noninteracting or with a contact interaction, namely

$$V(\mathbf{r} - \mathbf{r}') = g\,\delta(\mathbf{r} - \mathbf{r}') \,, \tag{8.37}$$

where $\delta(x)$ is the Dirac delta function and g the strength of the interaction. We adopt the path integral formalism, where the atomic bosons are described by the complex field $\psi(\mathbf{r}, \tau)$.

The Euclidean Lagrangian density of the free system in a D-dimensional box of volume L^D and with chemical potential μ is then given by

$$\mathcal{L} = \psi^* \left[\hbar \frac{\partial}{\partial \tau} - \frac{\hbar^2}{2m} \nabla^2 - \mu \right] \psi + \frac{1}{2} g\,|\psi|^4 \,. \tag{8.38}$$

Notice that this Lagrangian density is equivalent to the one of Eq. (8.35) because they give the same Euler-Lagrange equations.

We have seen that the partition function $\mathcal{Z}$ of the system at temperature T can then be written as

$$\mathcal{Z} = \int \mathcal{D}[\psi, \psi^*] \, \exp \left\{ -\frac{S[\psi, \psi^*]}{\hbar} \right\} \,, \tag{8.39}$$

where

$$S[\psi, \psi^*] = \int_0^{\hbar\beta} d\tau \int_{L^D} d^D\mathbf{r} \, \mathcal{L}(\psi, \psi^*) \tag{8.40}$$

is the Euclidean action and $\beta \equiv 1/(k_B T)$ with k_B being Boltzmann's constant.

The grand potential Ω of the system, which is a function of μ, T and g, is then obtained as

$$\Omega = -\frac{1}{\beta} \ln \mathcal{Z} \,. \tag{8.41}$$

We work in the superfluid phase where the global U(1) gauge symmetry of the system is spontaneously broken. For this reason we set

$$\psi(\mathbf{r}, \tau) = \psi_0 + \eta(\mathbf{r}, \tau) \,, \tag{8.42}$$

where $\eta(\mathbf{r}, \tau)$ is the complex field of bosonic fluctuations around the order parameter ψ_0 (condensate in 3D or quasi-condensate in 1D and 2D) of the system. We suppose that ψ_0 is constant in time, uniform in space and real.

8.2.1 Ideal Bose Gas: Critical Temperature and Condensate Fraction

First we analyze the case with $g = 0$, where exact analytical results can be obtained in any spatial dimension D. In this case the Lagrangian density reads

$$\mathcal{L} = -\mu \, \psi_0^2 + \eta^*(\mathbf{r}, \tau) \left[\hbar \frac{\partial}{\partial \tau} - \frac{\hbar^2}{2m} \nabla^2 - \mu \right] \eta(\mathbf{r}, \tau) . \tag{8.43}$$

Introducing the Fourier representation of the fluctuation field

$$\eta(\mathbf{r}, \tau) = \frac{1}{\sqrt{\hbar \beta L^D}} \sum_Q \eta_Q \, e^{i(\mathbf{q} \cdot \mathbf{r} - \omega \tau)} , \tag{8.44}$$

where $Q = (\mathbf{q}, i\omega_n)$ is the $D+1$ vector denoting the momenta $\mathbf{q}$ and frequencies $\omega_n = 2\pi n/(\beta\hbar)$ (which are usually called bosonic Matsubara frequencies), the Euclidean action of the ideal Bose gas can be written in a diagonal form as

$$S[\psi, \psi^*] = -\mu \, \psi_0^2 \, \hbar\beta \, L^D + \sum_Q \hbar \lambda_Q \, \eta_Q^* \eta_Q \tag{8.45}$$

with

$$\lambda_Q = \beta \left(-i \hbar \omega_n + \frac{\hbar^2 q^2}{2m} - \mu \right) . \tag{8.46}$$

Here $\eta_Q = \eta_{\mathbf{q}, i\omega_n} = \eta_{\mathbf{q}, n}$ is the Fourier transform of $\eta(\mathbf{r}, \tau)$ and the quantization of frequencies ω_n is a consequence of the periodicity $\eta(\mathbf{r}, \tau + \hbar\beta) = \eta(\mathbf{r}, \tau)$.

Taking into account that

$$\mathcal{Z} = \int \mathcal{D}[\eta, \eta^*] \exp \left\{ -\sum_Q \lambda_Q \eta_Q^* \eta_Q \right\} = \frac{1}{\prod_Q \lambda_Q} \tag{8.47}$$

and

$$\sum_{n=-\infty}^{+\infty} \ln(-i\,n + a) = \frac{1}{2} \sum_{n=-\infty}^{+\infty} \ln(n^2 + a^2) , \tag{8.48}$$

one finds the grand potential

$$\Omega = -\mu \, \psi_0^2 L^D + \frac{1}{\beta} \sum_Q \ln(\lambda_Q)$$

$$= -\mu \, \psi_0^2 L^D + \frac{1}{2\beta} \sum_{\mathbf{q}} \sum_{n=-\infty}^{+\infty} \ln[\beta^2(\hbar^2\omega_n^2 + \xi_q^2)] , \tag{8.49}$$

where ξ_q is the shifted free-particle spectrum, i.e.

$$\xi_q = \frac{\hbar^2 q^2}{2m} - \mu .$$ (8.50)

The sum over bosonic Matsubara frequencies gives

$$\frac{1}{2\beta} \sum_{n=-\infty}^{+\infty} \ln [\beta^2 (\hbar^2 \omega_n^2 + \xi_q^2)] = \frac{\xi_q}{2} + \frac{1}{\beta} \ln (1 - e^{-\beta \xi_q}) .$$ (8.51)

In fact, the derivative with respect to ξ_q of Eq. (8.51) reads

$$\frac{1}{\beta} \sum_{n=-\infty}^{+\infty} \frac{\xi_q}{\omega_n^2 + \xi_q^2} = \frac{1}{2} + \frac{1}{e^{\beta \xi_q} - 1} = \frac{1}{2} \coth \left(\frac{\beta \xi_q}{2}\right) ,$$ (8.52)

and this formula is a straightforward consequence of the exact result

$$\sum_{n=-\infty}^{+\infty} \frac{1}{n^2 + a^2} = \frac{\pi}{a} \coth (\pi a)$$ (8.53)

remembering that the bosonic Matsubara frequencies are given by $\omega_n = 2\pi n/\beta$.
   The grand potential finally reads

$$\Omega = \Omega_0 + \Omega^{(0)} + \Omega^{(T)} ,$$ (8.54)

where

$$\Omega_0 = -\mu \psi_0^2 L^D$$ (8.55)

is the grand potential of the order parameter,

$$\Omega^{(0)} = \frac{1}{2} \sum_{\mathbf{q}} \xi_q$$ (8.56)

is the zero-point energy of bosonic single-particle excitations, i.e. the zero-temperature contribution of quantum fluctuations, and

$$\Omega^{(T)} = \frac{1}{\beta} \sum_{\mathbf{q}} \ln \left(1 - e^{-\beta \xi_q}\right)$$ (8.57)

takes into account thermal fluctuations.

In the continuum limit, where $\sum_{\mathbf{q}} \rightarrow L^D \int d^D \mathbf{q}/(2\pi)^D$, the zero-point energy

$$\frac{\Omega^{(0)}}{L^D} = \frac{1}{2} \frac{S_D}{(2\pi)^D} \int_0^{+\infty} dq \, q^{D-1} \left(\frac{\hbar^2 q^2}{2m} - \mu \right) \tag{8.58}$$

of the ideal Bose gas is clearly ultraviolet divergent at any integer dimension D, i.e. for $D = 1, 2, 3$. Here $S_D = 2\pi^{D/2}/\Gamma(D/2)$ is the solid angle in D dimensions with $\Gamma(x)$ the Euler gamma function. We shall show that this divergent zero-point energy of the ideal Bose gas is completely eliminated by dimensional regularization. Consequently, the exact grand potential of the ideal Bose gas is given by

$$\frac{\Omega}{L^D} = -\mu \, \psi_0^2 + \frac{1}{\beta L^D} \sum_{\mathbf{q}} \ln \left(1 - e^{-\beta \xi_q} \right) . \tag{8.59}$$

We notice that ψ_0 is not a free parameter but must be determined by minimizing Ω_0, namely

$$\left(\frac{\partial \Omega_0}{\partial \psi_0} \right)_{\mu, T, L^D} = 0 , \tag{8.60}$$

from which one finds that

$$\psi_0 = \begin{cases} 0 & \text{if } \mu \neq 0 \\ \text{any value} & \text{if } \mu = 0 \end{cases} \tag{8.61}$$

The number density $n = N/L^D$ is obtained from the thermodynamic relation

$$n = -\frac{1}{L^D} \left(\frac{\partial \Omega}{\partial \mu} \right)_{T, L^D, \psi_0} , \tag{8.62}$$

which gives:

$$n = \psi_0^2 + \frac{1}{L^D} \sum_{\mathbf{q}} \frac{1}{e^{\beta \xi_q} - 1} . \tag{8.63}$$

In the continuum limit $\sum_{\mathbf{q}} \rightarrow L^D \int d^D \mathbf{q}/(2\pi)^D$, by setting $\mu = 0$ (condensed phase) from Eq. (8.63) one gets

$$n = n_0 + \int \frac{d^D \mathbf{q}}{(2\pi)^D} \frac{1}{e^{\frac{\hbar^2 q^2}{2 m k_B T}} - 1} , \tag{8.64}$$

which gives the Bose-Einstein condensate density $n_0 = \psi_0^2$ as a function of the temperature T.

The critical temperature T_c of Bose-Einstein condensation is obtained setting $\psi_0 = 0$ in the previous equation. In this way one finds

$$k_B T_c = \begin{cases} \frac{1}{2\pi\zeta(3/2)^{2/3}} \frac{\hbar^2}{m} n^{2/3} & \text{for } D = 3 \\ 0 & \text{for } D = 2 \\ \text{no solution} & \text{for } D = 1 \end{cases} \tag{8.65}$$

where $\zeta(x)$ is the Riemann zeta function. This result was extended to interacting systems by David Mermin and Herbert Wagner in 1966. The so-called Mermin-Wagner theorem states that there is no Bose-Einstein condensation at finite temperature in homogeneous systems with sufficiently short-range interactions in dimensions $D \leq 2$.

In the three-dimensional case ($D = 3$) from the equation

$$n = n_0 + \int \frac{d^3\mathbf{q}}{(2\pi)^3} \frac{1}{e^{\frac{\hbar^2 q^2}{2mk_B T}} - 1} , \tag{8.66}$$

we find

$$n = n_0 + \zeta(3/2) \left(\frac{mk_B T}{2\pi\hbar^2} \right)^{3/2} . \tag{8.67}$$

It follows that

$$\frac{n_0}{n} = 1 - \frac{\zeta(3/2) \left(\frac{mk_B}{2\pi\hbar^2} \right)^{3/2} T^{3/2}}{n} = 1 - \frac{\zeta(3/2) \left(\frac{mk_B}{2\pi\hbar^2} \right)^{3/2} T^{3/2}}{\zeta(3/2) \left(\frac{mk_B}{2\pi\hbar^2} \right)^{3/2} T_c^{3/2}} . \tag{8.68}$$

Thus, the condensate fraction reads

$$\frac{n_0}{n} = 1 - \left(\frac{T}{T_c} \right)^{3/2} . \tag{8.69}$$

The critical temperature of Eq. (8.65) and this formula for the condensate fraction of non-interacting massive bosons were obtained in 1925 by Albert Einstein extending previous results derived by Satyendra Nath Bose for a gas a photons (massless bosons) (Fig. 8.1).

8.2.2 Interacting Bose Gas: Bogoliubov Spectrum

Let us now consider a system of bosons with a repulsive contact interaction, i.e. with $g > 0$ in

$$\mathcal{L} = \psi^* \left[\hbar \frac{\partial}{\partial \tau} - \frac{\hbar^2}{2m} \nabla^2 - \mu \right] \psi + \frac{1}{2} g |\psi|^4 . \tag{8.70}$$

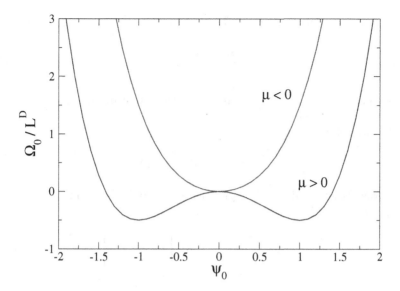

Fig. 8.1 Mean-field grand potential Ω_0 as a function of the real order parameter ψ_0 for an interacting Bose gas, see Eq. (8.72)

In this case one finds immediately the partition function of the uniform and constant order parameter ψ_0 as

$$\mathcal{Z}_0 = \exp\left\{-\frac{S_0}{\hbar}\right\} = \exp\left\{-\beta\,\Omega_0\right\}, \tag{8.71}$$

where $S_0 = S[\psi_0]$ and the grand potential Ω_0 reads

$$\frac{\Omega_0}{L^D} = -\mu\,\psi_0^2 + \frac{1}{2}\,g\,\psi_0^4. \tag{8.72}$$

Again, the constant, uniform and real order parameter ψ_0 is obtained by minimizing Ω_0 as

$$\left(\frac{\partial\Omega_0}{\partial\psi_0}\right)_{\mu,T,L^D} = 0, \tag{8.73}$$

from which one finds the relation between order parameter and chemical potential

$$\mu = g\,\psi_0^2. \tag{8.74}$$

showing that in the superfluid broken phase the chemical potential is positive and

$$\psi_0 = \sqrt{\frac{\mu}{g}}, \tag{8.75}$$

Inserting this relation into Eq. (8.72) we find

$$\frac{\Omega_0}{L^D} = -\frac{\mu^2}{2g} . \tag{8.76}$$

Clearly, this equation of state is lacking important informations encoded in quantum and thermal fluctuations.

Thus we consider again

$$\psi(\mathbf{r}, \tau) = \psi_0 + \eta(\mathbf{r}, \tau) \tag{8.77}$$

and expand the action $S[\psi, \psi^*]$ around ψ_0 up to quadratic (Gaussian) order in $\eta(\mathbf{r}, \tau)$ and $\bar{\eta}(\mathbf{r}, \tau)$. We find

$$Z = Z_0 \int \mathcal{D}[\eta, \eta^*] \exp\left\{-\frac{S_g[\eta, \eta^*]}{\hbar}\right\} , \tag{8.78}$$

where

$$S_g[\eta, \eta^*] = \frac{1}{2} \sum_Q (\eta_Q^*, \eta_{-Q}) \, \mathbf{M}_Q \begin{pmatrix} \eta_Q \\ \eta_{-Q}^* \end{pmatrix} \tag{8.79}$$

is the Gaussian action of fluctuations in reciprocal space with $Q = (\mathbf{q}, i\omega_n)$ the $D+1$ vector denoting the momenta $\mathbf{q}$ and bosonic Matsubara frequencies $\omega_n = 2\pi n/(\beta\hbar)$, and

$$\mathbf{M}_Q = \beta \begin{pmatrix} -i\hbar\omega_n + \frac{\hbar^2 q^2}{2m} - \mu + 2g\psi_0^2 & -g\psi_0^2 \\ -g\psi_0^2 & i\hbar\omega_n + \frac{\hbar^2 q^2}{2m} - \mu + 2g\psi_0^2 \end{pmatrix} \tag{8.80}$$

is the inverse fluctuation propagator.

Integrating over the bosonic fields $\eta(Q)$ and $\bar{\eta}(Q)$ in Eq. (8.78) one finds the Gaussian grand potential

$$\begin{aligned}
\Omega_g &= \frac{1}{2\beta} \sum_Q \ln \text{Det}(\mathbf{M}_Q) \\
&= \frac{1}{2\beta} \sum_{\mathbf{q}} \sum_{n=-\infty}^{+\infty} \ln[\beta^2(\hbar^2\omega_n^2 + E_q^2)] ,
\end{aligned} \tag{8.81}$$

where E_q is given by

$$E_q = \sqrt{\left(\frac{\hbar^2 q^2}{2m} - \mu + 2g\psi_0^2\right)^2 - g^2\psi_0^4} . \tag{8.82}$$

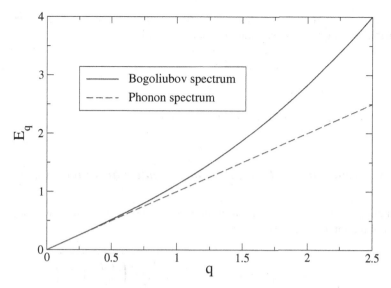

Fig. 8.2 Bogoliubov spectrum, given by Eq. (8.83), and its low-momentum phonon spectrum $E_q = c_B \hbar q$, where $c_B = \sqrt{\mu/m}$ is the sound velocity. Energy E_q in units of μ and momentum q in units of $\sqrt{m\mu/\hbar^2}$

By using $\psi_0 = \sqrt{\mu/g}$ the spectrum becomes

$$E_q = \sqrt{\frac{\hbar^2 q^2}{2m}\left(\frac{\hbar^2 q^2}{2m} + 2\mu\right)} \simeq c_B \hbar q \quad \text{for small } q , \tag{8.83}$$

which is the familiar Bogoliubov spectrum, with $c_B = \sqrt{\mu/m}$ (Fig. 8.2).

Again, the sum over bosonic Matsubara frequencies gives

$$\frac{1}{2\beta} \sum_{n=-\infty}^{+\infty} \ln[\beta^2(\hbar^2\omega_n^2 + E_q^2)] = \frac{E_q}{2} + \frac{1}{\beta}\ln(1 - e^{-\beta E_q}) . \tag{8.84}$$

The total grand potential may then be written as

$$\Omega = \Omega_0 + \Omega_g^{(0)} + \Omega_g^{(T)} , \tag{8.85}$$

where Ω_0 is given by Eq. (8.76).

$$\Omega_g^{(0)} = \frac{1}{2}\sum_{\mathbf{q}} E_q \tag{8.86}$$

is the zero-point energy of bosonic collective excitations, i.e. the zero-temperature contribution of quantum Gaussian fluctuations, while

$$\Omega_g^{(T)} = \frac{1}{\beta} \sum_{\mathbf{q}} \ln \left(1 - e^{-\beta E_q} \right) \tag{8.87}$$

takes into account thermal Gaussian fluctuations.

8.2.3 Dimensional Regularization of Gaussian Fluctuations

We notice that the continuum limit of the zero-point energy for the Gaussian fluctuation of the interacting Bose gas

$$\frac{\Omega_g^{(0)}}{L^D} = \frac{1}{2} \frac{S_D}{(2\pi)^D} \int_0^{+\infty} dq \, q^{D-1} \sqrt{\frac{\hbar^2 q^2}{2m} \left(\frac{\hbar^2 q^2}{2m} + 2\mu \right)} \tag{8.88}$$

is ultraviolet divergent at any integer dimension D.

We may rewrite Eq. (8.88) for the zero point energy of a repulsive Bose gas in dimension D as:

$$\frac{\Omega_g^{(0)}}{L^D} = \frac{S_D (2\mu)^{\frac{D}{2}+1}}{4(2\pi)^D} \left(\frac{2m}{\hbar^2} \right)^{\frac{D}{2}} B\left(\frac{D+1}{2}, -\frac{D+2}{2} \right), \tag{8.89}$$

where $B(x, y)$ is the Euler Beta function

$$B(x, y) = \int_0^{+\infty} dt \frac{t^{x-1}}{(1+t)^{x+y}}, \qquad Re(x), Re(y) > 0 \tag{8.90}$$

which may be continued to complex values of x and y as

$$B(x, y) = \frac{\Gamma(x) \, \Gamma(y)}{\Gamma(x+y)} . \tag{8.91}$$

We rewrite Eq. (8.88) using Eq. (8.91) to get

$$\frac{\Omega_g^{(0)}}{L^D} = \frac{S_D (2\mu)^{\frac{D}{2}+1}}{4(2\pi)^D} \left(\frac{2m}{\hbar^2} \right)^{\frac{D}{2}} \frac{\Gamma(\frac{D+1}{2}) \Gamma(-\frac{D+2}{2})}{\Gamma(-\frac{1}{2})} . \tag{8.92}$$

This expression is now finite when $D = 1$ or $D = 3$, while it is still divergent if $D = 2$ since $\Gamma(p)$ diverges for integers $p \leq 0$. In particular, setting $D = 3$ in Eq. (8.92) we get

$$\frac{\Omega_g^{(0)}}{L^3} = \frac{8}{15\pi^2}(\frac{m}{\hbar^2})^{3/2}\mu^{5/2} \, .$$ (8.93)

In conclusion, the total grand potential of the three-dimensional Bose gas is then given by

$$\frac{\Omega}{L^3} = -\frac{\mu^2}{2\,g} + \frac{8}{15\pi^2}(\frac{m}{\hbar^2})^{3/2}\mu^{5/2} + \frac{1}{\beta L^3}\sum_q \ln\left(1 - e^{-\beta E_q}\right) \, .$$ (8.94)

This result was obtained for the first time in 1957 by Tsung-Dao Lee, Kerson Huang, and Chen Ning Yang without invoking dimensional regularization.

8.3 Gross-Pitaevskii Equation

Let us consider again a system of bosons with a repulsive contact interaction, i.e. let us set $g > 0$ in

$$\mathcal{L} = \psi^*\left[\hbar\frac{\partial}{\partial\tau} - \frac{\hbar^2}{2m}\nabla^2 - \mu + U(\mathbf{r})\right]\psi + \frac{1}{2}\,g\,|\psi|^4 \, ,$$ (8.95)

where $U(\mathbf{r})$ is the external trapping potential.

In this inhomogeneous case the partition function of the mean-field (saddle-point) order parameter $\psi_0(\mathbf{r}, \tau)$ reads

$$\mathcal{Z}_0 = \exp\left\{-\frac{S_0}{\hbar}\right\} = \exp\left\{-\beta\,\Omega_0\right\} \, ,$$ (8.96)

where

$$S_0 = S[\psi_0(\mathbf{r}, \tau), \psi_0^*(\mathbf{r}, \tau)] = \int dt \int d^3\mathbf{r}\,\mathcal{L}(\psi_0, \psi_0^*)$$ (8.97)

and $\psi_0(\mathbf{r}, \tau)$ is the field $\psi(\mathbf{r}, \tau)$ which minimizes $S[\psi(\mathbf{r}, \tau)]$, namely such that

$$\delta S[\psi(\mathbf{r}, \tau), \psi^*(\mathbf{r}, \tau)] = 0 \, .$$ (8.98)

The minimization of the action S gives the Euler-Lagrange equation

$$\frac{\partial\mathcal{L}}{\partial\psi^*} - \frac{\partial}{\partial\tau}\frac{\partial\mathcal{L}}{\partial(\frac{\partial}{\partial\tau}\psi^*)} - \nabla\frac{\partial\mathcal{L}}{\partial(\nabla\psi^*)} + \nabla^2\frac{\partial\mathcal{L}}{\partial(\nabla^2\psi^*)} = 0$$ (8.99)

from which we get

$$\hbar \frac{\partial}{\partial \tau} \psi_0(\mathbf{r}, \tau) = \left[-\frac{\hbar^2}{2m} \nabla^2 - \mu + U(\mathbf{r}) + g \, |\psi_0(\mathbf{r}, \tau)|^2 \right] \psi_0(\mathbf{r}, \tau) \, . \qquad (8.100)$$

This is the Gross-Pitaevskii equation with imaginary time τ. Moreover, at the mean-field level, the number of bosons reads

$$N = -\frac{\partial \Omega_0}{\partial \mu} = \int d^3 \mathbf{r} \, |\psi_0(\mathbf{r}, \tau)|^2 \, . \qquad (8.101)$$

Going back to real time $t = i\tau$ we obtain

$$i\hbar \frac{\partial}{\partial t} \psi_0(\mathbf{r}, t) = \left[-\frac{\hbar^2}{2m} \nabla^2 - \mu + U(\mathbf{r}) + g \, |\psi_0(\mathbf{r}, t)|^2 \right] \psi_0(\mathbf{r}, t) \, , \qquad (8.102)$$

that is the real-time (time-dependent) Gross-Pitaevskii equation, with the normalization condition

$$N = \int d^3 \mathbf{r} \, |\psi_0(\mathbf{r}, t)|^2 \, . \qquad (8.103)$$

Immediately one deduces the corresponding stationary Gross-Pitaevskii equation

$$\left[-\frac{\hbar^2}{2m} \nabla^2 + U(\mathbf{r}) + g \, |\psi_0(\mathbf{r})|^2 \right] \psi_0(\mathbf{r}) = \mu \, \psi_0(\mathbf{r}) \, . \qquad (8.104)$$

8.3.1 Superfluid Hydrodynamics

Quite remarkably the zero-temperature time-dependent Gross-Pitaevskii equation (8.102) can be rewritten as the equations of superfluid hydrodynamics. In fact, setting

$$\psi_0(\mathbf{r}, t) = n(\mathbf{r}, t)^{1/2} \, e^{i\theta(\mathbf{r}, t)} \, , \qquad (8.105)$$

and inserting this formula into Eq. (8.102) one finds

$$\frac{\partial}{\partial t} n + \nabla \cdot (n \, \mathbf{v}) = 0 \, , \qquad (8.106)$$

$$m \frac{\partial}{\partial t} \mathbf{v} + \nabla \left[\frac{1}{2} m v^2 + U(\mathbf{r}) + g n - \frac{\hbar^2}{2m} \frac{\nabla^2 \sqrt{n}}{\sqrt{n}} \right] = 0 \, , \qquad (8.107)$$

where

$$\mathbf{v}(\mathbf{r}, t) = \frac{\hbar}{m} \nabla \theta(\mathbf{r}, t) \qquad (8.108)$$

is the local velocity field, that is (by definition) irrotational, i.e. such that

$$\mathbf{\nabla} \wedge \mathbf{v} = \mathbf{0} \ . \tag{8.109}$$

Equations (8.106) and (8.107) are, respectively, the equation of continuity and the equation of conservation of linear momentum for a irrotational and inviscid fluid. The zero-temperature equation of state of this superfluid, i.e. the local chemical potential as a function of the local density and its derivatives, can be written as

$$\mu(n, \nabla^2 n) = gn - \frac{\hbar^2}{2m} \frac{\nabla^2 \sqrt{n}}{\sqrt{n}} \ , \tag{8.110}$$

where the second term, which is usually called quantum pressure, becomes negligible in the high-density regime. In 1941 Lev Landau, on the basis of previous important theoretical hints of Fritz London and Laszlo Tisza, developed the two-fluid theory of superfluid ^{4}He. At zero temperature this two-fluid theory gives Eqs. (8.106) and (8.107).

We observe that quantum effects are encoded not only in this equation of state but also into the properties of the local field $\mathbf{v}(\mathbf{r}, t)$: it is proportional to the gradient of a scalar field, $\theta(\mathbf{r}, t)$, that is the angle of the phase of the single-valued complex wavefunction $\psi_0(\mathbf{r}, t)$. Consequently, $\mathbf{v}(\mathbf{r}, t)$ satisfies the equation

$$\oint_C \mathbf{v} \cdot d\mathbf{r} = 2\pi \frac{\hbar}{m} k \tag{8.111}$$

for any closed contour C, with k an integer number. In other words, the circulation is quantized in units of $\hbar/m$, and this property is strictly related to the existence of quantized vortices.

For the sake of completeness we observe that, in general, for a rotational and viscous fluid with quite generic zero-temperature equation of state $\mu(n, \nabla n, \nabla^2 n, ...)$ the equations of hydrodynamics are given by

$$\frac{\partial}{\partial t} n + \mathbf{\nabla} \cdot (n \, \mathbf{v}) = 0 \tag{8.112}$$

$$m\frac{\partial}{\partial t} \mathbf{v} - \eta \nabla^2 \mathbf{v} + \mathbf{\nabla} \left[\frac{1}{2} m v^2 + U(\mathbf{r}) + \mu(n, \nabla n, \nabla^2 n, ...) \right] = m\mathbf{v} \wedge (\mathbf{\nabla} \wedge \mathbf{v}) \tag{8.113}$$

where η is the viscosity and a rotational term appears on the right side of Eq. (8.113). These equations are called zero-temperature Navier-Stokes equations. Any generic fluid (in the collisional regime) satisfies these equations. Also a superfluid is described by Eqs. (8.112) and (8.113), but only if $\eta = 0$ and under the additional constraints given by Eqs. (8.109) and (8.111).

Let us consider Eqs. (8.106) and (8.107) assuming that $U(\mathbf{r}) = 0$. We set

$$n(\mathbf{r}, t) = n_{eq} + \delta n(\mathbf{r}, t) \,, \tag{8.114}$$

$$\mathbf{v}(\mathbf{r}, t) = \mathbf{0} + \delta \mathbf{v}(\mathbf{r}, t) \,, \tag{8.115}$$

where $\delta n(\mathbf{r}, t)$ and $\delta \mathbf{v}(\mathbf{r}, t)$ represent small variations with respect to the uniform and constant stationary configuration n_{eq}. In this way, neglecting quadratic terms in the variations (linearization) from Eqs. (8.106) and (8.107) we get the linear equations of motion

$$\frac{\partial}{\partial t} \delta n + n_{eq} \nabla \cdot \delta \mathbf{v} = 0 \,, \tag{8.116}$$

$$\frac{\partial}{\partial t} \delta \mathbf{v} + \frac{c_s^2}{n_{eq}} \nabla \delta n - \frac{\hbar^2}{4m^2 n_{eq}} \nabla (\nabla^2 \delta n) = \mathbf{0} \,, \tag{8.117}$$

where c_s is the sound velocity of the bosonic superfluid, given by

$$mc_s^2 = g \, n_{eq} \,. \tag{8.118}$$

The linear equations of motion can be arranged in the form of the following wave equation

$$\left[\frac{\partial^2}{\partial t^2} - c_s^2 \nabla^2 + \frac{\hbar^2}{4m^2} \nabla^4 \right] \delta n(\mathbf{r}, t) = 0 \,. \tag{8.119}$$

This wave equation admits monochromatic plane-wave solutions, where the frequency ω and the wave vector $\mathbf{q}$ are related by the dispersion formula $\omega = \omega(q)$ given by

$$\hbar \omega(q) = \sqrt{\frac{\hbar^2 q^2}{2m} \left(\frac{\hbar^2 q^2}{2m} + 2mc_s^2 \right)} \,. \tag{8.120}$$

We have thus obtained, in a different way, the Bogoliubov spectrum of elementary excitations.

8.3.2 Bright and Dark Solitons

The real-time Gross-Pitaevskii equation (8.102) can be seen as the Euler-Lagrange equation associated to the minimization of the following (real-time) action functional

$$S = \int dt \, d^3\mathbf{r} \left\{ \psi_0^*(\mathbf{r}, t) \left[i\hbar \frac{\partial}{\partial t} + \frac{\hbar^2}{2m} \nabla^2 - U(\mathbf{r}) \right] \psi_0(\mathbf{r}, t) - \frac{1}{2} g \, |\psi_0(\mathbf{r}, t)|^4 \right\} \,. \tag{8.121}$$

Let us consider a very strong harmonic confinement of frequency $\omega_\perp$ along x and y but no confinement along z, namely

$$U(\mathbf{r}) = \frac{1}{2}m\omega_\perp^2(x^2 + y^2) . \tag{8.122}$$

We adopt the variational ansatz

$$\psi_0(\mathbf{r}, t) = f(z, t)\frac{1}{\pi^{1/2}a_\perp^2} \exp\left(\frac{x^2 + y^2}{2a_\perp^2}\right) , \tag{8.123}$$

where $f(z, t)$ is the axial wave function and $a_\perp = \sqrt{\hbar/(m\omega_\perp)}$ is the characteristic length of the transverse harmonic confinement. By inserting Eq. (8.123) into Eq. (8.121) and integrating along x and y, the resulting effective action functional depends on the field $f(z, t)$. One easily finds that the Euler-Lagrange equation of the axial wavefunction $f(z, t)$ reads

$$i\hbar\frac{\partial}{\partial t}f(z, t) = \left[-\frac{\hbar^2}{2m}\frac{\partial^2}{\partial z^2} + g_{1D}|f(z, t)|^2\right]f(z, t) , \tag{8.124}$$

where

$$g_{1D} = \frac{g}{2\pi a_\perp^2} \tag{8.125}$$

is the effective one-dimensional interaction strength and the additive constant $\hbar\omega_\perp$ has been omitted because it does not affect the dynamics.

Equation (8.124) is a one-dimensional cubic nonlinear Schrödinger equation. In 1972 Vladimir Zakharov and Aleksei Shabat found that this equation admits solitonic (i.e. shape invariant) analytical solutions. In particular, for $g_{1D} < 0$ (self-focusing nonlinearity) one finds the localized bright soliton solution

$$f(z, t) = \sqrt{\frac{m|g_{1D}|}{8\hbar^2}}\, Sech\left[\frac{m|g_{1D}|}{4\hbar^2}(z - vt)\right]e^{iv(z-vt)}e^{i(mv^2/2-\mu)t/\hbar} , \tag{8.126}$$

where $Sech[x]$ is the hyperbolic secant, v is the arbitrary velocity of propagation of the shape-invariant bright soliton and $\mu = -mg_{1D}^2/(16\hbar^2)$. Instead, for $g_{1D} > 0$ (self-defocusing nonlinearity) one finds the localized but not normalized dark soliton solution

$$f(z, t) = \sqrt{\frac{g_{1D}\phi_\infty^2 - mv^2}{g_{1D}}}\left(Tanh\left[(z - vt)\sqrt{g_{1D}\phi_\infty^2 - v^2}\right] + i\sqrt{\frac{mv^2}{g_{1D}\phi_\infty^2 - mv^2}}\right)e^{-i\mu t/\hbar} , \tag{8.127}$$

where $Tanh[x]$ is the hyperbolic tangent, $\phi_\infty > 0$ is the constant value of the field amplitude $|f(z, t)|$ at infinity ($z \to \pm\infty$), and v is the velocity of propagation of the dark soliton. The parameter $\mu = g_{1D}\phi_\infty^2$ is obtained by the asymptotic behavior of the field $f(z, t)$. For $0 < |v| < c_s$, where $c_s = \sqrt{g\phi_\infty^2}$ is the axial sound velocity, the minimum value of the field $f(z, t)$ is greater than zero and the wave is called gray soliton. If $v = 0$ (stationary dark soliton) the minimum of the field $f(z, t)$ is zero and the wave is called black soliton. The velocity c_s is the maximal velocity of the dark soliton, namely the velocity of a wave of infinitesimal amplitude, i.e. a sound wave, propagating in a medium of density $\rho_\infty = \phi_\infty^2$.

8.4   Solved Problems

Problem 8.1

Consider a Bose-Einstein condensate made of N ultracold atoms with mass m and scattering length a_s, under external harmonic confinement of frequency ω. Neglecting the kinetic term in the Gross-Pitaevskii equation derive the density profile $n(\mathbf{r})$ and the chemical potential μ of the system.

Solution

The stationary Gross-Pitaevskii equation of the Bose-Einstein condensate is given by

$$\left[-\frac{\hbar^2}{2m}\nabla^2 + U(\mathbf{r}) + g\,|\psi_0(\mathbf{r})|^2 \right] \psi_0(\mathbf{r}) = \mu\,\psi_0(\mathbf{r}) ,$$

where

$$U(\mathbf{r}) = \frac{1}{2}m\omega^2(x^2 + y^2 + z^2)$$

and

$$g = \frac{4\pi\hbar^2}{m}a_s .$$

Neglecting the kinetic term (second derivatives) in the Gross-Pitaevskii equation one immediately finds the density profile

$$n(\mathbf{r}) = |\psi_0(\mathbf{r})|^2 = \frac{1}{g}\,(\mu - U(\mathbf{r}))\,\Theta\,(\mu - U(\mathbf{r}))$$

with $\Theta(x)$ the Heaviside step function. Imposing the condition

$$N = \int d^3\mathbf{r}\,n(\mathbf{r})$$

which fixes the chemical potential μ, after integration one gets

$$\mu = \frac{1}{2}\hbar\omega\left(\frac{15Na_s}{a_H}\right)^{2/5}$$

with $a_H = \sqrt{\hbar/(m\omega)}$ the characteristic length of the isotropic harmonic potential.

Problem 8.2

Consider a bosonic gas of N non-interacting spinless particles under harmonic confinement given by the external trapping potential

$$U(\mathbf{r}) = \frac{1}{2}m\omega^2(x^2 + y^2 + z^2) .$$

Calculate the critical temperature T_c of Bose-Einstein condensation as a function of the number N of bosons and the condensate fraction N_0/N as a function of the temperature T.

Solution

One must generalize Eq. (8.63) taking into account the external potential $U(\mathbf{r})$ acting on bosons. This can be done introducing the shifted single-particle energy

$$\xi_q(\mathbf{r}) = \frac{\hbar^2 q^2}{2m} + U(\mathbf{r}) - \mu .$$

In this way one can write the local number density as

$$n(\mathbf{r}) = n_0(\mathbf{r}) + \int \frac{d^3\mathbf{q}}{(2\pi)^3} \frac{1}{e^{\frac{\xi_q(\mathbf{r})}{k_BT}} - 1} ,$$

where $n_0(\mathbf{r})$ is the local condensate density. Below T_c one can set $\mu = 0$. Moreover, at T_c by definition $n_0(\mathbf{r}) = 0$ and consequently

$$n(\mathbf{r}) = \int \frac{d^3\mathbf{q}}{(2\pi)^3} \frac{1}{e^{\frac{\frac{\hbar^2 q^2}{2m} + U(\mathbf{r})}{k_B T_c}} - 1} .$$

After integration over momenta one obtains

$$n(\mathbf{r}) = \frac{1}{\lambda_{T_c}^3} g_{3/2}(e^{-\frac{U(\mathbf{r})}{k_B T_c}}) ,$$

where

$$\lambda_{T_c} = \left(\frac{2\pi\hbar^2}{mk_B T_c}\right)^{1/2}$$

is the de Broglie wavelength at the critical temperature T_c and $g_n(x)$ is the Bose function, defined as

$$g_n(x) = \frac{1}{\Gamma(n)} \int_0^{+\infty} dy \frac{x\, y^{n-1}\, e^{-y}}{1 - x\, e^{-y}} = \sum_{i=1}^{+\infty} \frac{x^i}{i^n} ,$$

with $\Gamma(n)$ the Euler gamma function, and the last equality holds only if $|x| < 1$. Moreover, integrating over coordinates one gets

$$N = \int d^3r\, n(\mathbf{r}) = g_3(1) \left(\frac{\hbar\omega}{k_B T_c} \right)^3 ,$$

from which

$$T_c = \frac{\hbar\omega}{g_3(1)\, k_B} N^{1/3} .$$

For $T \leq T_c$ one can write

$$N = N_0 + \int \frac{d^3r\, d^3q}{(2\pi)^3} \frac{1}{e^{\frac{\frac{\hbar^2 q^2}{2m} + U(\mathbf{r})}{k_B T}} - 1}$$

and consequently

$$\frac{N_0}{N} = 1 - \frac{1}{N} \int \frac{d^3r\, d^3q}{(2\pi)^3} \frac{1}{e^{\frac{\frac{\hbar^2 q^2}{2m} + U(\mathbf{r})}{k_B T}} - 1} .$$

After integration over coordinates and momenta, and taking into account the formula of T_c, one finally gets

$$\frac{N_0}{N} = 1 - \left(\frac{T}{T_c} \right)^3 .$$

Problem 8.3
Calculate, at the Gaussian level, the zero-temperature grand potential Ω of a 1D uniform Bose gas with repulsive 1D strength g_{1D} by using dimensional regularization.

Solution
At the Gaussian level the zero-temperature grand potential can be written as

$$\Omega = \Omega_0 + \Omega_g^{(0)} ,$$

where

$$\Omega_0 = -\frac{\mu^2}{2g_{1D}}$$

is the mean-field grand potential with μ the chemical potential. Within the dimensional regularization, in a D-dimensional space, the beyond-mean-field contribution $\Omega_g^{(0)}$ is instead given by Eq. (8.89). Setting $D = 1$ one immediately finds

$$\Omega_g^{(0)} = -\frac{2}{3\pi} \left(\frac{m}{\hbar^2}\right)^{1/2} \mu^{1/2} \, .$$

Problem 8.4

Given the 1D stationary Gross-Pitaevskii equation

$$\left[-\frac{\hbar^2}{2m}\frac{d^2}{dz^2} + g_{1D}|f(z)|^2\right] f(z) = \mu \, f(z) \, ,$$

with g_{1D} the attractive (negative) strength, prove by construction that it admits the stationary solution

$$f(z) = \sqrt{\frac{m|g_{1D}|}{8\hbar^2}} \, \text{Sech}\left[\frac{m|g_{1D}|}{4\hbar^2}z\right]$$

with

$$\mu = -\frac{m \, g_{1D}^2}{16 \, \hbar^2} \, .$$

Solution

Let us assume that $f(z)$ is real. Then the 1D stationary Gross-Pitaevskii equation can be rewritten as

$$f''(z) = -\frac{\partial W(f)}{\partial f} \, ,$$

where

$$W(f) = \frac{1}{2}\frac{m|g_{1D}|}{\hbar^2} f^4 + \frac{m\mu}{\hbar^2} f^2 \, .$$

Thus, $f(z)$ can be seen as the "coordinate" for a fictitious particle at "time" z. The constant of motion of the problem reads

$$K = \frac{1}{2}f'(z)^2 + W(f) \, ,$$

from which one finds

$$\frac{df}{dz} = \sqrt{K - W(f)} \, .$$

Imposing that $f(z) \to 0$ as $|z| \to \infty$ one gets $K = 0$ and consequently

$$\frac{df}{\sqrt{-W(f)}} = dz \, ,$$

or explicity

$$\frac{df}{\sqrt{-\frac{1}{2}\frac{m|g_{1D}|}{\hbar^2}f^4 + \frac{m|\mu|}{\hbar^2}f^2}} = dz \,,$$

with $\mu < 0$. Inserting the integrals one obtains

$$\int_{f(0)}^{f(z)} \frac{df}{\sqrt{-\frac{1}{2}\frac{m|g_{1D}|}{\hbar^2}f^4 + \frac{m|\mu|}{\hbar^2}f^2}} = z$$

Setting $f'(0) = 0$, from the definition of K and using $K = 0$ one finds $W(f(0)) = 0$ and therefore

$$f(0) = \sqrt{\frac{2|\mu|}{|g_{1D}|}} \,.$$

After integration

$$\frac{1}{\sqrt{\frac{m|\mu|}{\hbar}}} ArcSech\left[\sqrt{\frac{|g_{1D}|}{2|\mu|}} f(z)\right] = z$$

from which

$$f(z) = \sqrt{\frac{2|\mu|}{|g_{1D}|}} Sech\left[\sqrt{\frac{m|\mu|}{\hbar^2}} z\right] \,.$$

Imposing the normalization condition

$$\int dz\, f(z)^2 = 1 \,,$$

one obtains

$$\mu = -\frac{m\, g_{1D}^2}{16\, \hbar^2} \,.$$

Further Reading

For the functional integration of the non-relativistic bosonic field:
N. Nagaosa, Quantum Field Theory in Condensed Matter Physics, Chap. 1, Sects. 1.1 and 1.2 (Springer, Berlin, 1999).
H.T.C. Stoof, K.B. Gubbels, and D.B.M. Dickerscheid, Ultracold Quantum Fields, Chap. 6, Sects. 6.1, 6.2, 6.3, and 6.4 (Springer, Berlin, 2009).
A. Altland and B. Simons, Condensed Matter Field Theory, Chap. 2, Sect. 2.1 (Cambridge Univ. Press, Cambridge, 2006).

For a review of methods of regularization in the case of interacting bosons:

L. Salasnich and F. Toigo, Zero-point energy of ultracold atoms, Phys. Rep. **640**, 1 (2016).

For bright and dark solitons in the nonlinear Schrödinger equation:

A. Scott (ed.), Encyclopedia of Nonlinear Science (Taylor and Francis, New York, 2005).

Chapter 9
Functional Integration for the Fermionic Field

In this chapter we consider the quantum statistical mechanics of identical fermions by introducing the formalism of functional integration for the fermionic field. We then derive the equation of state of non-interacting and weakly-interacting repulsive fermions. We also discuss the paring of electrons within the mean-field Bardeen-Cooper-Schrieffer (BCS) theory of low-temperature superconductivity and the Ginzburg-Landau theory, which is quite successful near the critical temperature. The BCS theory is then extended to investigate the BCS-BEC crossover of ultracold atoms from the weak-coupling BCS regime of Cooper pairs to the strong-coupling regime where composite bosons form giving rise to Bose-Einstein condensation (BEC) of molecules.

9.1 Partition Function of Interacting Fermions

For a system of non-relativistic interacting two-spin-component fermions the partition function is given by

$$\mathcal{Z} = Tr[e^{-\beta(\hat{H} - \mu_\uparrow \hat{N}_\uparrow - \mu_\downarrow \hat{N}_\downarrow)}] , \qquad (9.1)$$

where the quantum-field-theory Hamiltonian is

$$\hat{H} = \sum_{\sigma=\uparrow,\downarrow} \int d^3\mathbf{r} \; \hat{\psi}_\sigma^+(\mathbf{r}) \left(-\frac{\hbar^2}{2m} \nabla^2 + U_\sigma(\mathbf{r}) \right) \hat{\psi}_\sigma(\mathbf{r})$$
$$+ \frac{1}{2} \sum_{\sigma,\sigma'=\uparrow,\downarrow} \int d^3\mathbf{r} \, d^3\mathbf{r}' \; \hat{\psi}_\sigma^+(\mathbf{r}) \, \hat{\psi}_{\sigma'}^+(\mathbf{r}') \, V_{\sigma,\sigma'}(\mathbf{r}, \mathbf{r}') \, \hat{\psi}_{\sigma'}(\mathbf{r}') \, \hat{\psi}_\sigma(\mathbf{r}) , \qquad (9.2)$$

with $U_\sigma(\mathbf{r})$ the external potential acting on fermions with spin σ, $V_{\sigma,\sigma'}(\mathbf{r}, \mathbf{r}')$ the inter-particle interaction, while the number operator for fermions with spin σ reads

© Springer International Publishing AG 2017
L. Salasnich, *Quantum Physics of Light and Matter*, UNITEXT for Physics,
DOI 10.1007/978-3-319-52998-1_9

$$\hat{N}_\sigma = \int d^3\mathbf{r}\, \hat{\psi}_\sigma^+(\mathbf{r})\hat{\psi}_\sigma(\mathbf{r}) \ . \tag{9.3}$$

As usual, $\beta = 1/(k_B T)$ with k_B the Boltzmann constant and T the temperature, while μ_σ is the chemical potential for the fermions with spin σ.

The Hamiltonian (9.2) generalizes the one, valid for spinless fermions, described in Chap. 7. Assuming that (i) the external confinement does not depend on spin σ, (ii) there is a contact interaction of strength g between fermions, (iii) there is chemical equilibrium, i.e. $\mu_\uparrow = \mu_\downarrow = \mu$, we can write

$$\mathcal{Z} = Tr[e^{-\beta(\hat{H} - \mu\hat{N})}] \tag{9.4}$$

with

$$\hat{N} = \hat{N}_\uparrow + \hat{N}_\downarrow \tag{9.5}$$

and

$$\hat{H} = \sum_{\sigma=\uparrow,\downarrow} \int d^3\mathbf{r}\, \hat{\psi}_\sigma^+(\mathbf{r})\left(-\frac{\hbar^2}{2m}\nabla^2 + U(\mathbf{r})\right)\hat{\psi}_\sigma(\mathbf{r}) + \frac{g}{2}\int d^3\mathbf{r}\, \hat{\psi}_\uparrow^+(\mathbf{r})\,\hat{\psi}_\downarrow^+(\mathbf{r})\,\hat{\psi}_\downarrow(\mathbf{r})\,\hat{\psi}_\uparrow(\mathbf{r}) \ . \tag{9.6}$$

Notice that, as shown in Chap. 6, in the presence of a contact inter-particle potential, there is an interaction only between fermions with opposite spins.

In the previous chapter we have seen that the partition function of identical bosons can be expressed as a functional integral, taking into account that the trace of an operator can be performed by using any complete set of basis states, and in particular the basis of generalized coherent states of bosons. The same reasoning can be applied to the partition function of identical fermions by using the basis of generalized coherent states of fermions. As shown in Chap. 7, in the case of fermionic coherent states one must use complex Grassmann variables and complex Grassmann fields. Thus, we can write

$$\hat{\psi}_\sigma(\mathbf{r})\,|\psi\rangle = \psi_\sigma(\mathbf{r})\,|\psi\rangle \ , \tag{9.7}$$

where $\psi_\sigma(\mathbf{r})$ is the complex Grassmann field with spin σ and $|\psi\rangle$ is the generalized fermionic coherent state. The complex conjugate of the Grassmann field $\psi_\sigma(\mathbf{r})$ is denoted as $\bar{\psi}_\sigma(\mathbf{r})$ and it satisfies the equation

$$\langle\psi|\hat{\psi}_\sigma^+(\mathbf{r}) = \langle\psi|\bar{\psi}_\sigma(\mathbf{r}) \ . \tag{9.8}$$

By expanding the fermionic field operator $\hat{\psi}_\sigma(\mathbf{r})$ as

$$\hat{\psi}_\sigma(\mathbf{r}) = \sum_\alpha \hat{c}_{\alpha,\sigma}\,\phi_\alpha(\mathbf{r}) \ , \tag{9.9}$$

where $\phi_\alpha(\mathbf{r})$ are orthonormal eigenfunctions of single-particle states $|\alpha\rangle$ and $\hat{c}_{\alpha,\sigma}$ is the fermionic operator which annihilates a fermion with spin σ in the state $|\alpha\rangle$, we clearly have

$$\psi_\sigma(\mathbf{r}) = \sum_\alpha c_{\alpha,\sigma}\, \phi_\alpha(\mathbf{r}) \tag{9.10}$$

with

$$\hat{c}_{\alpha,\sigma}|\psi\rangle = c_{\alpha,\sigma}|\psi\rangle \quad \text{for any } \alpha \text{ and } \sigma = \uparrow, \downarrow \tag{9.11}$$

and $c_{\alpha,\sigma}$ the complex Grassmann variables for the single-particle state $|\alpha\rangle$ and spin σ.

Making a long story short, in full analogy with the bosonic case, the partition function of identical fermions described by the many-body Hamiltonian (9.6) can be written as

$$\mathcal{Z} = \int \mathcal{D}[\psi_\sigma, \bar{\psi}_\sigma] \exp\left\{ -\frac{1}{\hbar} S[\psi_\sigma, \bar{\psi}_\sigma] \right\}, \tag{9.12}$$

where

$$S[\psi_\sigma, \bar{\psi}_\sigma] = \int_0^{\hbar\beta} d\tau \int d^3\mathbf{r}\, \mathcal{L}(\psi_s, \bar{\psi}_s) \tag{9.13}$$

is the Euclidean action functional and

$$\mathcal{L} = \sum_{\sigma=\uparrow,\downarrow} \bar{\psi}_\sigma \left[\hbar\partial_\tau - \frac{\hbar^2}{2m}\nabla^2 + U(\mathbf{r}) - \mu \right]\psi_\sigma + g\,\bar{\psi}_\uparrow\,\bar{\psi}_\downarrow\,\psi_\downarrow\,\psi_\uparrow \tag{9.14}$$

is the Euclidean Lagrangian density, where $\psi_\sigma(\mathbf{r}, \tau)$ and $\bar{\psi}_\sigma(\mathbf{r}, \tau)$ are the Grassmann fields which depend also on the imaginary time τ. Notice that one picks up a minus sign each time one permutes two Gaussian fields. In the fermionic functional integral (9.12) the integration measure reads

$$\int \mathcal{D}[\psi_\sigma, \bar{\psi}_\sigma] = \int \prod_{\sigma=\uparrow,\downarrow} \prod_{(\mathbf{r},\tau)} d\bar{\psi}_\sigma(\mathbf{r}, \tau)\, d\psi_\sigma(\mathbf{r}, \tau), \tag{9.15}$$

and the boundary conditions for the imaginary time τ are anti-periodic, i.e. $\psi(\mathbf{r}, 0) = -\psi(\mathbf{r}, \hbar\beta)$.

9.2 Ideal Fermi Gas

We consider a D-dimensional ($D = 1, 2, 3$) Fermi gas of noninteracting fermions in a D-dimensional box of volume L^D. Under these conditions exact analytical results can be obtained in any spatial dimension D. In this case the Euclidean Lagrangian density is given by

$$\mathcal{L} = \sum_{\sigma=\uparrow,\downarrow} \bar{\psi}_\sigma \left[\hbar \partial_\tau - \frac{\hbar^2}{2m} \nabla^2 - \mu \right] \psi_\sigma , \tag{9.16}$$

and the Euclidean action can be written in a diagonal form as

$$S[\psi, \psi^*] = \sum_{\sigma=\uparrow,\downarrow} \sum_Q \hbar \lambda_Q \, \bar{\psi}_{Q,\sigma} \psi_{Q,\sigma} \tag{9.17}$$

where $Q = (\mathbf{q}, i\omega_n)$ is the $D+1$ vector denoting the momenta $\mathbf{q}$ and frequencies $\omega_n = 2\pi(n + 1/2)/(\beta\hbar)$ (which are usually called fermionic Matsubara frequencies), and

$$\lambda_Q = \beta \left(-i\hbar\omega_n + \frac{\hbar^2 q^2}{2m} - \mu \right) . \tag{9.18}$$

Here $\psi_{Q,\sigma} = \psi_{\mathbf{q},i\omega_n,\sigma} = \psi_{\mathbf{q},n,\sigma}$ is the Fourier transform of $\psi_\sigma(\mathbf{r}, \tau)$ and the quantization of frequencies ω_n is a consequence of the anti-periodicity $\psi_\sigma(\mathbf{r}, \tau + \hbar\beta) = -\psi_\sigma(\mathbf{r}, \tau)$.

Taking into account that

$$\mathcal{Z} = \int \mathcal{D}[\psi_\sigma, \bar{\psi}_\sigma] \exp\left\{ -\sum_{\sigma=\uparrow,\downarrow} \sum_Q \lambda_{Q,\sigma} \bar{\psi}_{Q,\sigma} \psi_{Q,\sigma} \right\} = \prod_{\sigma=\uparrow,\downarrow} \prod_Q \lambda_Q , \tag{9.19}$$

the grand potential reads

$$\Omega = -\frac{1}{\beta} \ln \mathcal{Z} = -\frac{2}{\beta} \sum_Q \ln(\lambda_Q) = -\frac{1}{\beta} \sum_{\mathbf{q}} \sum_{n=-\infty}^{+\infty} \ln[\beta^2(\hbar^2\omega_n^2 + \xi_q^2)] , \tag{9.20}$$

summing over the two spin states and with ξ_q the shifted free-particle spectrum, i.e.

$$\xi_q = \frac{\hbar^2 q^2}{2m} - \mu . \tag{9.21}$$

The Matsubara frequencies of fermions are given by $\omega_n = (2n+1)\pi/\beta$ due to the anti-symmetry of the Grassmann fields. In this way one gets the exact result

$$\frac{1}{\beta} \sum_{n=-\infty}^{+\infty} \ln[\beta^2(\hbar^2\omega_n^2 + \xi_q^2)] = \xi_q + \frac{2}{\beta} \ln(1 + e^{-\beta\xi_q}) . \tag{9.22}$$

Thus, the grand potential finally reads

$$\Omega = \Omega^{(0)} + \Omega^{(T)} , \tag{9.23}$$

where

$$\Omega^{(0)} = -\sum_q \xi_q \tag{9.24}$$

is the zero-point energy of fermionic single-particle excitations, i.e. the zero-temperature contribution of quantum fluctuations, and

$$\Omega^{(T)} = -\frac{2}{\beta} \sum_q \ln\left(1 + e^{-\beta \xi_q}\right) \tag{9.25}$$

takes into account thermal fluctuations.

In the continuum limit, where $\sum_q \to L^D \int d^D q/(2\pi)^D$, the zero-point energy

$$\frac{\Omega^{(0)}}{L^D} = -\frac{S_D}{(2\pi)^D} \int_0^{+\infty} dq\, q^{D-1} \left(\frac{\hbar^2 q^2}{2m} - \mu\right) \tag{9.26}$$

of the ideal Fermi gas is clearly ultraviolet divergent at any integer dimension D, i.e. for $D = 1, 2, 3$. Here $S_D = 2\pi^{D/2}/\Gamma(D/2)$ is the solid angle in D dimensions with $\Gamma(x)$ the Euler gamma function. This divergent zero-point energy of the ideal Fermi gas is completely eliminated by dimensional regularization. Consequently, the exact grand potential of the ideal Fermi gas is given by

$$\frac{\Omega}{L^D} = -\frac{2}{\beta} \int \frac{d^D q}{(2\pi)^D} \ln\left(1 + e^{-\beta \xi_q}\right), \tag{9.27}$$

and the corresponding number density reads

$$n = \frac{1}{L^D}\frac{\partial\Omega}{\partial\mu} = 2\int \frac{d^D q}{(2\pi)^D}\frac{1}{e^{\beta\xi_q} + 1}. \tag{9.28}$$

At zero temperature this formula becomes

$$n = \frac{1}{L^D}\frac{\partial\Omega}{\partial\mu} = 2\int \frac{d^D q}{(2\pi)^D}\Theta(-\xi_q), \tag{9.29}$$

where $\Theta(x)$ is the Heaviside step function. After integration one obtains

$$\mu = \begin{cases} \frac{\hbar^2}{2m}(3\pi^2 n)^{2/3} & \text{for } D = 3 \\ \frac{\hbar^2}{2m}2\pi n & \text{for } D = 2 \\ \frac{\hbar^2}{2m}\frac{\pi^2}{4}n^2 & \text{for } D = 1 \end{cases} \tag{9.30}$$

where the chemical potential μ can be identified with the Fermi energy ϵ_F.

9.3 Repulsive Fermions: Hartree-Fock Approximation

Let us now consider identical fermions with a repulsive contact interaction. This system is described by the Lagrangian density (9.14) with $g > 0$. In addition, we suppose that the system has the same number of spin up and spin down fermions, i.e. $n_\uparrow = n_\downarrow = n/2$, with n the total number density of fermions.

The problem cannot be solved exactly and we adopt a mean-field Hartee-Fock approximation for the quartic term, namely

$$g\,\bar\psi_\uparrow\,\bar\psi_\downarrow\,\psi_\downarrow\,\psi_\uparrow \simeq g\,\frac{n}{2}\bar\psi_\uparrow\psi_\uparrow + g\,\frac{n}{2}\bar\psi_\downarrow\psi_\downarrow - g\,\frac{n^2}{4}\,. \tag{9.31}$$

In this way the Lagrangian density (9.14) becomes quadratic

$$\mathcal{L} = \sum_{\sigma=\uparrow,\downarrow}\bar\psi_\sigma\left[\hbar\partial_\tau - \frac{\hbar^2}{2m}\nabla^2 - \mu + \frac{1}{2}gn\right]\psi_\sigma - g\,\frac{n^2}{4}\,, \tag{9.32}$$

By repeating the procedure of the previous section it is straightforward to show that the mean-field grand potential reads

$$\frac{\Omega}{L^D} = -\frac{2}{\beta}\int\frac{d^D\mathbf{q}}{(2\pi)^D}\ln\left(1 + e^{-\beta E_q}\right) - g\,\frac{n^2}{4}\,, \tag{9.33}$$

where the single-particle Hartree-Fock energy E_q given by

$$E_q = \frac{\hbar^2 q^2}{2m} - \mu + \frac{1}{2}gn\,. \tag{9.34}$$

The number density n must satisfy the equation

$$n = \frac{1}{L^D}\frac{\partial\Omega}{\partial\mu} = 2\int\frac{d^D\mathbf{q}}{(2\pi)^D}\frac{1}{e^{\beta E_q} + 1}\,, \tag{9.35}$$

and, at zero-temperature, after integration one obtains

$$\mu = \begin{cases} \frac{\hbar^2}{2m}(3\pi^2 n)^{2/3} + \frac{1}{2}gn & \text{for } D = 3 \\ \frac{\hbar^2}{2m}2\pi n + \frac{1}{2}gn & \text{for } D = 2 \\ \frac{\hbar^2}{2m}\frac{\pi^2}{4}n^2 + \frac{1}{2}gn & \text{for } D = 1 \end{cases} \tag{9.36}$$

In the three-dimensional case this result was obtained for the first time in 1957 by Kerson Huang and Chen Ning Yang.

9.4 Attractive Fermions: BCS Approximation

Let us now consider identical fermions with an attractive contact interaction. This system is described by the Lagrangian density (9.14) with $g < 0$. In addition, we suppose that the system has the same number of spin up and spin down fermions, i.e. $n_\uparrow = n_\downarrow = n/2$, with n the total number density of fermions.

As discussed in the previous section, the problem cannot be solved exactly due to the presence of the quartic term in the Lagrangian density. Also in this case we adopt a mean-field treatment which is however quite different with respect to the Hartree-Fock one of repulsive fermions. We suppose that, due to the attractive interaction, fermions with opposite spin can create a sort of bosonic particle with spin zero.

In particular we use the following mean-field Bardeen-Cooper-Schriffer (BCS) approximation

$$- g_0 \, \bar{\psi}_\uparrow \, \bar{\psi}_\downarrow \, \psi_\downarrow \, \psi_\uparrow \simeq -\Delta^* \psi_\downarrow \psi_\uparrow - \Delta \bar{\psi}_\uparrow \bar{\psi}_\downarrow + \frac{|\Delta|^2}{g_0} \, , \qquad (9.37)$$

where $g_0 = -g > 0$ and Δ is a complex order parameter usually called pairing energy or pairing energy gap. For attractive fermions this BCS approximation gives much better agreement with exact calculations (and also with experimental data) than the Hartree-Fock approximation. In this way the Lagrangian density (9.14) becomes quadratic and given by

$$\mathcal{L} = \bar{\psi}_\sigma \left[\hbar \partial_\tau - \frac{\hbar^2}{2m} \nabla^2 - \mu \right] \psi_\sigma - \Delta^* \, \psi_\downarrow \, \psi_\uparrow - \Delta \bar{\psi}_\uparrow \, \bar{\psi}_\downarrow + \frac{|\Delta|^2}{g_0} \, . \qquad (9.38)$$

Given the BCS Lagrangian density (9.38), the corresponding BCS Euclidean action can be written in a diagonal form as

$$S[\psi, \psi^*] = \sum_{\sigma=\uparrow,\downarrow} \sum_Q \hbar \lambda_Q \, \bar{\psi}_{Q,\sigma} \psi_{Q,\sigma} + \frac{\Delta^2}{g_0} L^D \hbar \beta \, , \qquad (9.39)$$

where

$$\lambda_Q = \beta \, (-i \hbar \omega_n + E_q) \qquad (9.40)$$

with

$$E_q = \sqrt{ \left(\frac{\hbar^2 q^2}{2m} - \mu \right)^2 + |\Delta|^2 } \qquad (9.41)$$

the single-particle fermionic excitations.

The mean-field grand potential can then be written as

$$\Omega = \Omega_0 + \Omega^{(0)} + \Omega^{(T)} \qquad (9.42)$$

including the grand potential of the order parameter Δ

$$\Omega_0 = \frac{\Delta^2}{g_0} L^D , \qquad (9.43)$$

the zero-point energy of fermionic single-particle excitations

$$\Omega^{(0)} = -\sum_q E_q , \qquad (9.44)$$

and the finite-temperature grand potential of the fermionic single-particle excitations

$$\Omega^{(T)} = \frac{2}{\beta} \sum_q \ln \left(1 + e^{-\beta E_q}\right) . \qquad (9.45)$$

The energy gap Δ which appears in the fermionic energy spectrum (9.41) is obtained by imposing that

$$\frac{\partial \Omega}{\partial \Delta} = 0 , \qquad (9.46)$$

from which one gets

$$\frac{1}{g_0} = \frac{1}{L^D} \sum_q \frac{\tanh\left(\frac{\beta E_q}{2}\right)}{2E_q} \qquad (9.47)$$

that is the so-called gap equation. The number density n is instead obtained from

$$n = -\frac{1}{L^D} \frac{\partial \Omega}{\partial \mu} , \qquad (9.48)$$

which gives the so-called number equation

$$n = \frac{1}{L^D} \sum_q \left(1 - \frac{\frac{\hbar^2 q^2 - \mu}{2m}}{E_q}\right) \tanh\left(\frac{\beta E_q}{2}\right) . \qquad (9.49)$$

In three-dimensions ($D = 3$) the gap equation (9.47) is ultraviolet divergent. It becomes convergent by introducing an ultraviolet cutoff Λ such that

$$\frac{1}{g_0} = \frac{1}{L^3} \sum_{|q|<\Lambda} \frac{\tanh\left(\frac{\beta E_q}{2}\right)}{2E_q} . \qquad (9.50)$$

For electrons in superconducting metals the cutoff energy $\hbar^2 \Lambda^2/(2m)$ can be identified with the Debye energy $\hbar \omega_D$, where ω_D is the Debye frequency, that is the maximum frequency of vibration for the atoms that make up the ionic crystal of the metal. The cutoff Λ is then quite natural because it accounts for the fact that phonons of ionic vibrations, which mediate the attraction between condition electrons, can at most transfer an energy $\hbar\omega_D$.

In the thermodynamic limit, where the volume L^3 goes to infinity, $\sum_{\mathbf{q}}$ can be replaced by $L^3 \int d^3\mathbf{q}/(2\pi)^3 = L^3 \int N(\xi)d\xi$ with $N(\xi) = \int d^3\mathbf{q}/(2\pi)^3 \, \delta(\xi - \xi_q)$, $\xi_q = \epsilon_q - \mu$, and $\epsilon_q = \hbar^2 q^2/(2m)$. In metals the condition $\hbar\omega_D \ll \mu$ is always satisfied, consequently we can use the approximation $\int N(\xi)d\xi \simeq N(0) \int d\xi$, where

$$N(0) = \int \frac{d^3\mathbf{q}}{(2\pi)^3} \, \delta(\mu - \epsilon_q) \tag{9.51}$$

is the density of states at the Fermi surface. In this way the gap equation (9.50) becomes

$$\frac{1}{g_0 N(0)} = \int_0^{\hbar\omega_D} \frac{\tanh(\frac{\beta}{2}\sqrt{\xi^2 + \Delta^2})}{\sqrt{\xi^2 + \Delta^2}} \, d\xi . \tag{9.52}$$

At zero-temperature, from Eq. (9.52) we find immediately a simple formula for the zero temperature energy gap $\Delta(0)$:

$$\frac{1}{g_0 N(0)} = \ln\left(\frac{\hbar\omega_D}{\Delta(0)} + \sqrt{1 + \frac{\hbar^2\omega_D^2}{\Delta(0)^2}}\right) . \tag{9.53}$$

Under the condition $\Delta(0) \ll \hbar\omega_D$, from Eq. (9.53) we find the weak-coupling BCS result

$$\Delta(0) = 2\hbar\omega_D \, \exp(-\frac{1}{g_0 N(0)}) \tag{9.54}$$

for the zero-temperature energy gap. Clearly the energy gap Δ decreases by increasing the temperature T and it becomes zero at the critical temperature T_c of the superconducting transition. Setting $\Delta = 0$ in Eq. (9.52) one finds

$$k_B T_c = 1.13 \, \hbar\omega_D \, \exp(-\frac{1}{g_0 N(0)}) . \tag{9.55}$$

that is the BCS critical temperature obtained by John Bardeen, Leon Cooper and John Robert Schrieffer in 1957. Remarkably, taking into account Eqs. (9.54) and (9.55), the BCS theory predicts that the zero-temperature energy gap $\Delta(0)$ of superconductors in the weak-coupling regime is related to the critical temperature T_c by the simple equation

$$\Delta(0) = 1.764 \, k_B T_c . \tag{9.56}$$

9.4.1 Condensate Fraction of Cooper Pairs

In the mean-field BCS approximation we have introduced the energy gap Δ. In general, this pairing order parameter is defined as

$$\Delta(\mathbf{r}, \tau) = g_0 \langle \psi_\downarrow(\mathbf{r}, \tau) \psi_\uparrow(\mathbf{r}, \tau) \rangle . \tag{9.57}$$

We have seen that a nonzero value of this quantity implies the formation of Cooper pairs (two fermions with opposite spins) which give rise to a quasi-bosonic particles of spin zero. As suggested by Chen Ning Yang in 1962, the condensate wavefunction of these Cooper pairs is given by

$$\Xi(\mathbf{r}, \mathbf{r}', \tau) = \langle \psi_\downarrow(\mathbf{r}, \tau) \psi_\uparrow(\mathbf{r}', \tau) \rangle . \tag{9.58}$$

This two-particle wave function is strictly related to the largest eigenvalue N_0 of two-body density matrix of the system. N_0 gives the number of Fermi pairs (Cooper pairs) in the lowest state, i.e. the condensate number of Fermi pairs, and it can be written as

$$N_0 = \int |\Xi(\mathbf{r}, \mathbf{r}', \tau)|^2 \, d^3\mathbf{r} \, d^3\mathbf{r}' . \tag{9.59}$$

A finite value for the condensate fraction $f = N_0/(N/2)$ in the thermodynamic limit $N \to \infty$ implies the so-called off-diagonal-long-range-order.

By using the formalism of functional integration, the condensate wavefunction of Cooper pairs can be written as

$$\Xi(\mathbf{r}, \mathbf{r}', \tau) = \frac{1}{Z} \int \mathcal{D}[\psi_\sigma, \bar{\psi}_\sigma] \, \psi_\downarrow(\mathbf{r}, \tau) \, \psi_\uparrow(\mathbf{r}', \tau) \, \exp\left\{ -\frac{1}{\hbar} S[\psi_\sigma, \bar{\psi}_\sigma] \right\} . \tag{9.60}$$

Inserting the BCS Euclidean action (9.39), after some straightforward but lengthy calculations one finds that the condensate number of Cooper pairs satisfies this expression

$$N_0 = \sum_q \frac{\Delta^2}{4E_q^2} \tanh^2\left(\frac{\beta E_q}{2}\right) . \tag{9.61}$$

For the electrons in a superconducting metal, taking into account Eq. (9.51) one then easily finds

$$n_0 = \frac{1}{2} N(0) \Delta^2 \int_0^{\hbar\omega_D} \frac{\tanh^2\left(\frac{\beta}{2}\sqrt{\xi^2 + \Delta^2}\right)}{\xi^2 + \Delta^2} \, d\xi \tag{9.62}$$

where $n_0 = N_0/L^3$ is the density of electrons in the condensate. At zero temperature Eq. (9.62) gives

$$n_0(0) = \frac{1}{2} N(0) \Delta(0) \arctan\left(\frac{\hbar\omega_D}{\Delta(0)}\right) . \tag{9.63}$$

This expression shows that the condensate density $n(0)$ can be expressed in terms of density of states $N(0)$, energy gap $\Delta(0)$ and Debye energy $\hbar\omega_D$. Consequently, the zero-temperature condensate fraction $f(0) = n_0(0)/(n/2)$ is given by

$$f(0) = \frac{1}{2}\frac{N(0)}{n}\Delta(0)\arctan(\frac{\hbar\omega_D}{\Delta(0)}) , \tag{9.64}$$

where n is the density of conduction electrons. Moreover, under the condition $\Delta(0) \ll \hbar\omega_D$, the condensate density can be written as

$$n_0(0) = \frac{\pi}{4}N(0)\Delta(0) . \tag{9.65}$$

We stress that in many real superconductors the simple mean-field BCS approximation is not accurate, and one has to take into account the retarded electron-electron interaction via phonons and also the Coulomb repulsion. The results obtained above by using the mean-field BCS theory are reliable in the weak-coupling regime, i.e. for $\Delta(0) \ll \hbar\omega_D$, where $gN(0) \leq 0.3$. Therefore we will continue our analysis of the condensate fraction only for a class of superconductors which satisfy this condition.

Within the free-electrons Sommerfeld approximation, the energy spectrum ϵ_q of conduction electrons has a simple quadratic behavior $\epsilon_q = \hbar^2 q^2/(2m^*)$ but m^*, the effective mass of electrons, does not coincide with the physical mass m. The free particle density of states $N_{free}(0)$ is related to the total density of conduction electrons by $n = 4N_{free}(0)\mu/3$ and the zero-temperature condensate fraction reads $f(0) = 3\pi\Delta(0)/(8\mu)$. To get a better estimate, we correct the free electron value of $N(0)$ by an effective mass as obtained from specific heat measurements, i.e. we use the expression $N(0) = (m^*/m)N_{free}(0)$.

In the first two columns of Table 9.1 we show the zero-temperature condensate density n_0 and condensate fraction $f(0)$ of simple metals obtained from Eqs. (9.63) and (9.64) by using experimental data of $\Delta(0)$ and ω_D. In the third column we report

Table 9.1 BCS predictions for weak-coupling superconducting metals: n_0 is the zero-temperature condensate density, $f(0) = n_0(0)/(n/2)$ is the zero-temperature condensate fraction, $gN(0)$ is the electron-phonon strength, and T_c is the critical temperature. Experimental data taken from N.W. Ashcroft and N.D. Mermin, Solid State Physics (Holt-Sounders, Philadelphia, 1976) and J.P. Carbotte, Rev. Mod. Phys. **62**, 1027 (1990)

	$n_0(0)$ [10^{-33} m^{-3}]	$f(0)$ [10^{-5}]	$gN(0)$	T_c [K]
Cd	4.18	0.9	0.179	0.51
Zn	7.72	1.2	0.172	0.79
Al	22.4	2.5	0.168	1.15
Tl	31.0	5.9	0.263	2.43
In	62.1	10.8	0.267	3.46
Sn	62.5	8.4	0.254	3.68

the electron-phonon strength $gN(0)$ calculated with Eq. (9.53) knowing $f(0)$. The table shows that indeed these simple metals are all in the weak-coupling regime. For completeness, in the fourth column we insert the theoretical determination (see Eq. (9.55)) of the critical temperature T_c, which is very reliable for these simple metals, when compared with the experimental data.

Let us now consider the behavior of the condensate density n_0 at finite temperature T. Under the condition $\hbar \omega_D \gg k_B T_c$, from Eq. (9.52) one finds that near T_c the energy gap goes to zero according to the power law

$$\Delta(T) = 3.06 \, k_B T_c \left(1 - \frac{T}{T_c}\right)^{1/2} . \tag{9.66}$$

Instead for the condensate density $n_0(T)$, from Eq. (9.62) and the previous expression, we find near T_c

$$n_0(T) = 0.43 \, N(0) \frac{\Delta(T)^2}{k_B T_c} = 4.03 \, N(0) k_B T_c \left(1 - \frac{T}{T_c}\right) . \tag{9.67}$$

Taking into account Eq. (9.64), the BCS theory predicts that the zero-temperature condensate density in superconductors in the weak-coupling regime, can be written as

$$n_0(0) = 1.39 \, N(0) k_B T_c . \tag{9.68}$$

9.4.2 Ginzburg-Landau Theory

In 1950, seven years before the BCS theory, Lev Landau and Vitaly Ginzburg proposed a phenomenological approach to describe the superconducting phase transition. The main idea is that, close to the critical temperature, the grand potential of a superconducting material can be written as

$$\Omega = \Omega_n + \Omega_s , \tag{9.69}$$

where Ω_n is the contribution due to the normal component and Ω_s is the contribution due to the emergence of a superconducting complex order parameter $\Psi(\mathbf{r})$ below a critical temperature. Within the Ginzburg-Landau approach, for a D-dimensional system of area L^D, the super component Ω_s is given by

$$\Omega_s = \int_{L^D} d^D \mathbf{r} \left\{ a(T) \, |\Psi(\mathbf{r})|^2 + \frac{b}{2} \, |\Psi(\mathbf{r})|^4 + \gamma \, |\nabla \Psi(\mathbf{r})|^2 \right\} , \tag{9.70}$$

where

$$a(T) = \alpha \, k_B \, (T - T_c) \tag{9.71}$$

is a parameter which depends on the temperature T (k_B is the Boltzmann constant) and becomes zero at the mean-field critical temperature T_c, while $b > 0$ and $\gamma > 0$ are temperature-independent phenomenological parameters.

The partition function Z of the system is given by

$$Z = e^{-\beta \Omega_n} Z_s \qquad (9.72)$$

where

$$Z_s = \int d[\Psi^*, \Psi] e^{-\beta \Omega_s[\Psi, \Psi^*]} , \qquad (9.73)$$

with $\beta = 1/(k_B T)$, and the thermal average of an observable A that is a functional of $\Psi(\mathbf{r})$ and $\Psi^*(\mathbf{r})$ reads

$$\langle A \rangle = \frac{1}{Z_s} \int d[\Psi, \Psi^*] A[\Psi, \Psi^*] e^{-\beta \Omega_s[\Psi, \Psi^*]} . \qquad (9.74)$$

Notice that in these functional integrals the complex field $\Psi(\mathbf{r})$ does not depend on time. As shown in the previous chapter, they are fully reliable at high temperatures where quantum fluctuations are negligible.

Assuming a real and uniform order parameter, i.e.

$$\Psi(\mathbf{r}) = \Psi_0 , \qquad (9.75)$$

the energy functional (9.70) with Eqs. (9.71) and (9.75) becomes

$$\frac{\Omega_{s0}}{L^D} = a(T) \Psi_0^2 + \frac{b}{2} \Psi_0^4 . \qquad (9.76)$$

Minimizing Ω_{s0} with respect to Ψ_0 one immediately finds

$$a(T) \Psi_0 + b \Psi_0^3 = 0 , \qquad (9.77)$$

and consequently

$$\Psi_0 = \begin{cases} 0 & \text{for } T \geq T_c \\ \sqrt{-\frac{a(T)}{b}} & \text{for } T < T_c \end{cases} . \qquad (9.78)$$

Thus, the uniform order parameter Ψ_0 becomes different from zero only below the mean-field critical temperature T_c, where, by definition, T_c is such that

$$a(T_c) = 0 . \qquad (9.79)$$

In 1959 Lev Gor'kov showed that the phenomenological Ginzburg-Landau theory can be deduced from the microscopic BCS theory. In particular, the order parameter Ψ_0 can be identified with the BCS energy gap Δ, while the coefficients α, b and γ of the Ginzburg-Landau energy functional (9.70) are directly related to the parameters of the BCS Lagrangian density (9.38), taking into account Eqs. (9.54), (9.55), (9.71), and (9.78). We notice, however, that the Ginzburg-Landau theory is somehow better than the the mean-field BCS theory because, in general, the Ginzburg-Landau order parameter $\Psi(\mathbf{r})$ is not uniform, while the mean-field BEC energy gap Δ is assumed to be uniform.

Let us now consider the effects of a space-dependent Ginzburg-Landau order parameter $\Psi(\mathbf{r})$. Extremizing the functional (9.70) with respect to $\psi^*(\mathbf{r})$ one gets the Euler-Lagrange equation

$$a(T)\,\psi + b\,|\Psi|^2\Psi - \gamma\,\nabla^2\Psi = 0\ . \tag{9.80}$$

We write the space-dependent order parameter $\Psi(\mathbf{r})$ in the following way

$$\Psi(\mathbf{r}) = \Phi_0 + \eta(\mathbf{r})\ , \tag{9.81}$$

where $\eta(\mathbf{r})$ represents a fluctuation with respect to the real and uniform configuration Φ_0 with the condition

$$\langle\eta\rangle = \langle\eta^*\rangle = 0\ , \tag{9.82}$$

where $\langle\cdot\rangle$ is the thermal average. Inserting Eq. (9.81) into Eq. (9.80), we find

$$a(T)\,\Phi_0 + b\,\Phi_0^3 + a(T)\,\eta + 2b\,\Phi_0^2\eta + b\,\Phi_0^2\eta^* + 2b\,\Phi_0|\eta|^2$$
$$+ b\,\Phi_0\eta^2 + b\,|\eta|^2\eta - \gamma\nabla^2\eta = 0\ , \tag{9.83}$$

and after thermal averaging we obtain

$$\left[a(T) + 2b\,\langle|\eta|^2\rangle + b\,\langle\eta^2\rangle\right]\Phi_0 + b\,\Phi_0^3 + b\,\langle|\eta|^2\eta\rangle = 0\ . \tag{9.84}$$

Clearly, only removing all thermal averages one recovers Eq. (3.4) and finds that Φ_0 is equal to Ψ_0 and given by Eq. (9.78).

We take into account thermal fluctuations of the order parameter by keeping $\langle|\eta|^2\rangle$ but neglecting the anomalous averages $\langle\eta^2\rangle$ and $\langle|\eta|^2\eta\rangle$. In this way we get

$$\left(a(T) + 2b\,\langle|\eta|^2\rangle\right)\Phi_0 + b\,\Phi_0^3 = 0\ , \tag{9.85}$$

and consequently

$$\Phi_0 = \begin{cases} 0 & \text{for } T \geq T_{c,B} \\ \sqrt{-\dfrac{a(T)+2b\langle|\eta|^2\rangle}{b}} & \text{for } T < T_{c,B} \end{cases}\ . \tag{9.86}$$

In this case the uniform order parameter Φ_0 becomes different from zero only below the beyond-mean-field critical temperature $T_{c,B}$ given by the equation

$$a(T_{c,B}) + 2b\,\langle|\eta|^2\rangle_{c,B} = 0\ . \tag{9.87}$$

It is important to stress that, to treat self-consistently the fluctuating field $\eta(\mathbf{r})$, our approximation requires $|\eta|^2\eta \simeq 2\langle|\eta|^2\rangle\eta$ and $\eta^2 \simeq 0$ in Eq. (9.83), which becomes

$$a(T)\eta + 2b\,\phi_0^2\eta + 2b\,\langle|\eta|^2\rangle\eta + b\,\phi_0\eta^* - \gamma\nabla^2\eta = 0 , \qquad (9.88)$$

taking into account Eq. (9.85).

Let us work with $T \geq T_{c,B}$ where $\phi_0 = 0$. Then Eq. (9.88) reduces to

$$a(T)\eta + 2b\,\langle|\eta|^2\rangle\eta - \gamma\nabla^2\eta = 0 , \qquad (9.89)$$

which is the Euler-Lagrange equation of the Gaussian energy functional

$$\Omega_{sg} = \int_{L^D} d^D\mathbf{r}\left\{(a(T) + 2b\langle|\eta|^2\rangle)|\eta(\mathbf{r})|^2 + \gamma\,|\nabla\eta(\mathbf{r})|^2\right\} . \qquad (9.90)$$

Expanding the field $\eta(\mathbf{r})$ in a plane-wave basis, i.e.

$$\eta(\mathbf{r}) = \sum_k \psi_k \frac{e^{ik\cdot\mathbf{r}}}{L^{D/2}} , \qquad (9.91)$$

it is straightforward to find

$$\Omega_{sg} = \sum_k \left(a(T) + 2b\langle|\eta|^2\rangle + \gamma k^2\right)|\psi_k|^2 . \qquad (9.92)$$

The superconductive partition function Z_s of the system is then given by

$$Z_s = \int d[\eta^*, \eta]\,e^{-\beta\Omega_{sg}[\eta,\eta^*]} , \qquad (9.93)$$

where $\beta = 1/(k_B T)$ and the thermal average

$$\langle|\eta|^2\rangle = \frac{1}{Z_s} \int d[\eta^*, \eta]\,|\eta(\mathbf{r})|^2\,e^{-\beta\Omega_{sg}[\eta,\eta^*]} \qquad (9.94)$$

reads

$$\langle|\eta|^2\rangle = \frac{1}{L^D} \sum_k \frac{1}{\beta\,(a(T) + 2b\langle|\eta|^2\rangle + \gamma k^2)} . \qquad (9.95)$$

At $T = T_{c,B}$, where Eq. (9.87) holds, we get

$$\langle|\eta|^2\rangle_c = \frac{1}{L^D} \sum_k \frac{k_B T_{c,B}}{\gamma k^2} . \qquad (9.96)$$

In the continuum of momenta $\mathbf{k}$, where $\sum_{\mathbf{k}} = L^D \int d^D\mathbf{k}/(2\pi)^D$, this thermal average (9.96) goes to infinity, however, usually there are physical cutoffs which constrain the values of k.

In the two-dimensional case ($D = 2$), where fluctuations of the mean-field critical temperature are expected to be relevant, from Eq. (9.96) we find

$$\langle |\eta|^2 \rangle_c = \frac{k_B T_{c,B}}{2\pi\gamma} \ln \left(\frac{\Lambda}{k_0} \right) , \tag{9.97}$$

where k_0 is an infrared cutoff and Λ an ultraviolet cutoff. Then Eq. (9.87) gives

$$\frac{T_c - T_{c,B}}{T_{c,B}} = 4\, Gi \, \ln \left(\frac{\Lambda}{k_0} \right) \tag{9.98}$$

with

$$Gi = \frac{b}{4\pi\alpha\gamma} \tag{9.99}$$

the so-called Ginzburg-Levanyuk number. In addition, one can write $\Lambda = 1/(2\xi)$ with $\xi = (\gamma/(\alpha k_B T_{c,B}))^{1/2}$ the coherence length and $k_0 = (k_B T_{c,B} \alpha\, Gi/\gamma)^{1/2}$. It follows that $\Lambda/k_0 = 1/(4\, Gi)^{1/2}$ and consequently

$$\frac{T_c - T_{c,B}}{T_{c,B}} = 2\, Gi \, \ln \left(\frac{1}{4\, Gi} \right) , \tag{9.100}$$

that is a simple analytical formula for the shift of the critical temperature due to thermal fluctuations of the Ginzburg-Landau order parameter.

9.4.3 BCS-BEC Crossover

Scattering theory plays an essential role in the description of a three-dimensional attractive Fermi gas of atoms which undergoes BCS-BEC crossover, i.e. the crossover of a fermionic superfluid from weakly-bound BCS-like Cooper pairs to the Bose-Einstein condensation (BEC) of strongly-bound molecules.

In 1947 Bogoliubov found a regularization procedure relating the physical s-wave scattering length a_F of the actual interatomic potential to the strength g of the model contact interaction:

$$\frac{m}{4\pi\hbar^2 a_F} = \frac{1}{g} + \frac{1}{L^3} \sum_{\mathbf{q}} \frac{m}{\hbar^2 q^2} . \tag{9.101}$$

This equation may be easily deduced from the Lippman-Schwinger equation for the T-matrix $\hat{T}(E)$ of scattering theory. In the low energy limit, the Lippman-Schwinger equation renormalizes the coupling constant of a contact interaction $V(\mathbf{r}) = g\,\delta(\mathbf{r})$ as

$$\frac{1}{g_r(E)} \equiv \frac{1}{T(E+i0^+)} = \frac{1}{g} + \frac{1}{L^D}\sum_q \frac{m}{\hbar^2 q^2 - E - i0^+}, \qquad (9.102)$$

where $T(E) = T_{\mathbf{k},\mathbf{k}'}(E) \equiv \langle \mathbf{k}|\hat{T}(E)|\mathbf{k}'\rangle$ when $k = k'$ and $E = \hbar^2 k^2/m$.

At very low energy one may relate the s-wave scattering length a_F, a physical measurable quantity, to $T(E)$ according to

$$T(E+i0^+) = \begin{cases} \dfrac{4\pi\hbar^2}{m}\dfrac{a_F}{1+ia_F\sqrt{\frac{mE}{\hbar^2}}} & \text{for } D = 3 \\[3mm] -\dfrac{2\pi\hbar^2}{m}\dfrac{1}{\ln\left(a_B\sqrt{\frac{mE}{\hbar^2}}e^\gamma/2\right)+i\pi/2} & \text{for } D = 2 \\[3mm] -\dfrac{2\hbar^2}{m}\dfrac{i\sqrt{\frac{mE}{\hbar^2}}}{1+ia_F\sqrt{\frac{mE}{\hbar^2}}} & \text{for } D = 1 \end{cases} \qquad (9.103)$$

with $\gamma = 0.577\ldots$ the Euler-Mascheroni constant. In three dimensions (D = 3) one can safely set $E = 0$, so that $T(0+i0^+) = 4\pi\hbar^2 a_F/m$, and Eq. (9.101) is immediately recovered. In 2D and 1D one must use Eq. (9.102) with a finite value of E, while in 1D one gets $T(E+i0^+) = -2\hbar^2/(ma_F)$ under the condition $a_F\sqrt{mE/\hbar^2} \gg 1$.

As recalled above, the bare interaction strength g appearing in the Lagrangian density (9.38) is related to the physical s-wave scattering length a_F of fermions by

$$\frac{m}{4\pi\hbar^2 a_F} = \frac{1}{g} + \frac{1}{L^3}\sum_{|\mathbf{q}|<\Lambda} \frac{m}{\hbar^2 q^2}, \qquad (9.104)$$

where, as usual, the ultraviolet cutoff Λ is introduced to avoid the divergence of the second term on the right side. The low energy scattering length is negative for an attractive potential if no bound states are present, while it becomes positive when the interaction is so attractive as to admit a bound state. Equation (9.104) allows for the change of sign of a_F as the strength of the attractive potential becomes more and more negative. In fact, in the continuum limit $\sum_{\mathbf{k}} \to L^3\int d^3\mathbf{k}/(2\pi)^3$, after integration over momenta, it reads

$$\frac{m}{4\pi\hbar^2 a_F} = \frac{1}{g} + \frac{m}{2\pi^2\hbar^2}\Lambda. \qquad (9.105)$$

Therefore, in the weak-coupling BCS limit, where $g \to 0^-$, the first term on the r.h.s. of Eq. (9.105) dominates and $a_F = mg/(4\pi\hbar^2) \to 0^-$, while in the strong-coupling BEC limit, where $g \to -\infty$, the second term on the r.h.s. of Eq. (9.105) dominates and $a_F = \pi/(2\Lambda) \to 0^+$ when Λ is sent to infinity.

In addition, from Eq. (9.105) one finds that a bound state, that is a pole in the T-matrix of Eq. (9.102), is possible only if $g < -2\pi^2\hbar^2/(m\Lambda)$. See Fig. 9.1. In the presence of a bound state of energy $-\epsilon_B$, for the T-matrix one has

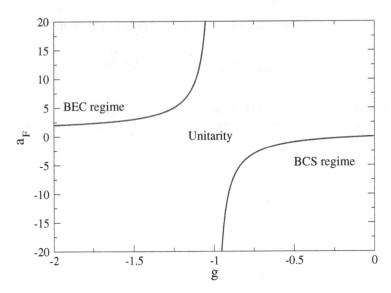

Fig. 9.1 Fermionic scattering length a_F as a function of the bare interaction strength g, for a finite ultraviolet cutoff Λ, see Eq. (9.105). In the plot a_F is in units of $m/(4\pi\hbar^2)$ and $m\Lambda/(2\pi^2\hbar^2) = 1$

$$\frac{1}{T(-\epsilon_B)} = 0 , \tag{9.106}$$

and consequently from Eq. (9.101) it follows

$$-\frac{1}{g} = \frac{1}{L^3} \sum_{|\mathbf{k}|<\Lambda} \frac{m}{\hbar^2 k^2 + m\epsilon_B} . \tag{9.107}$$

After integration over momenta one obtains

$$-\frac{1}{g} = \frac{m}{2\pi^2\hbar^2} \left(\Lambda - \sqrt{\frac{m\epsilon_B}{\hbar^2}} \arctan\left(\frac{\Lambda}{\sqrt{\frac{m\epsilon_B}{\hbar^2}}} \right) \right) , \tag{9.108}$$

and comparing Eq. (9.105) with Eq. (9.108) one finds

$$a_F = \sqrt{\frac{\hbar^2}{m\epsilon_B}} \frac{\pi}{2 \arctan\left(\Lambda\sqrt{\frac{\hbar^2}{m\epsilon_B}} \right)} . \tag{9.109}$$

In the limit $\Lambda \to +\infty$ from this interesting formula one gets

$$\epsilon_B = \frac{\hbar^2}{m a_F^2} , \qquad (9.110)$$

that is the familiar relation between the bound-state energy and the 3D s-wave scattering length a_F.

The full BCS-BEC crossover in 3D, which has been experimentally observed in 2004 with ultracold and dilute alkali-metal atoms, can be theoretically investigated by inserting Eq. (9.104) into Eq. (9.50). In this way, in the limit $\Lambda \to +\infty$, one gets

$$\frac{m}{4\pi\hbar^2 a_F} = \frac{1}{L^3} \sum_q \left(\frac{\tanh\left(\frac{\beta E_q}{2}\right)}{2 E_q} - \frac{m}{\hbar^2 q^2} \right) \qquad (9.111)$$

that is a renormalized gap equation valid both for $a_F < 0$ and $a_F > 0$. Solving this renormalized gap equation, coupled to the number equation (9.49), one finds simple analytical results for the chemical potential μ, the energy gap Δ, and the condensate fraction n_0/n (see Eq. 9.61) in limiting cases. In particular, introducing the adimensional strength $y = 1/(k_F a_F)$ with $k_F = (3\pi^2 n)^{1/3}$, three relevant limiting cases are: (i) the BCS regime, where $y \ll -1$, $\mu/\Delta \gg 1$ and the size of weakly bound Cooper pairs exceeds the typical interparticle spacing k_F^{-1}; (ii) the unitarity limit, with $y = 0$ (thus $a_F = \pm\infty$); (iii) the BEC regime, where $y \gg 1$, $\mu/\Delta \ll -1$ and the fermions condense as a gas of tight dimers. The main results are the following:

(i) *BCS regime:* μ approaches the non-interacting Fermi energy $\epsilon_F = \hbar^2 k_F^2/(2m)$. One finds that the condensate density is given by

$$n_0 \simeq \frac{m k_F}{8\pi\hbar^2} \Delta = \frac{3\pi}{2e^2} n \exp\left(\frac{\pi}{2} y\right) , \qquad (9.112)$$

with an exponentially small energy gap $\Delta = 8e^{-2} \epsilon_F \exp(y\pi/2)$.

(ii) *Unitarity limit:* one finds $\mu \simeq 0.59\epsilon_F$ and $\Delta \simeq 0.69\epsilon_F$, and the condensate fraction is

$$\frac{2 N_0}{N} \simeq 0.6994 . \qquad (9.113)$$

(iii) *BEC regime:* the condensate fraction approaches the ideal Bose gas value

$$\frac{2 N_0}{N} \simeq 1 , \qquad (9.114)$$

with all pairs (molecular dimers) moving into the condensate.

9.5 Solved Problems

Problem 9.1
Given two quantities A and B with averages $\langle A \rangle$ and $\langle B \rangle$, prove the mean-field result

$$A\,B \simeq \langle A \rangle\,B + A\,\langle B \rangle - \langle A \rangle\,\langle B \rangle$$

neglecting quadratic fluctuations.

Solution

First we set

$$A = \langle A \rangle + \delta_A$$
$$B = \langle B \rangle + \delta_B$$

where

$$\delta_A = A - \langle A \rangle$$
$$\delta_B = B - \langle B \rangle$$

are the fluctuations of A and B. Then we can write

$$A\,B = \Big(\langle A \rangle + \delta_A \Big)\Big(\langle B \rangle + \delta_B \Big) = \langle A \rangle\,\langle B \rangle + \langle A \rangle\,\delta_B + \langle B \rangle\,\delta_A + \delta_A\,\delta_B \; .$$

Neglecting the term $\delta_A\,\delta_B$ and taking into account the defintion of fluctuations, the previous equation becomes

$$
\begin{aligned}
A\,B &\simeq \langle A \rangle\,\langle B \rangle + \langle A \rangle\,\delta_B + \langle B \rangle\,\delta_A = \langle A \rangle\,\langle B \rangle + \langle A \rangle\,(B - \langle B \rangle) + \langle B \rangle\,(A - \langle A \rangle) \\
&= \langle A \rangle\,\langle B \rangle + \langle A \rangle\,B - \langle A \rangle\,\langle B \rangle + \langle B \rangle\,A - \langle B \rangle\,\langle A \rangle \\
&= \langle A \rangle\,B + A\,\langle B \rangle - \langle A \rangle\,\langle B \rangle \; ,
\end{aligned}
$$

by using the fact that averages commutes with themselves and also with quantities.

Problem 9.2
Consider a two-component three-dimensional ($D = 3$) Fermi gas with repulsive contact interaction of strength g. Within the mean-field Hartree-Fock approximation, find the critical strength g_c above which the zero-temperature uniform unpolarized configuration becomes unstable (3D Stoner instability).

Solution

Assuming that $\mu_\uparrow \neq \mu_\downarrow$ and $n_\uparrow \neq n_\downarrow$, the mean-field Hartee-Fock approximation for the quartic term gives

$$g \, \bar{\psi}_\uparrow \, \bar{\psi}_\downarrow \, \psi_\downarrow \, \psi_\uparrow \simeq g \, n_\downarrow \bar{\psi}_\uparrow \psi_\uparrow + g \, n_\uparrow \bar{\psi}_\downarrow \psi_\downarrow - g n_\uparrow n_\downarrow \, .$$

In this way the effective Lagrangian density is given by

$$\mathcal{L} = \sum_{\sigma=\uparrow,\downarrow} \bar{\psi}_\sigma \left[\hbar \partial_\tau - \frac{\hbar^2}{2m} \nabla^2 - \mu_\sigma + g n_\sigma \right] \psi_\sigma - g n_\uparrow n_\downarrow \, ,$$

and the mean-field grand potential reads

$$\frac{\Omega}{L^3} = -\frac{1}{\beta} \sum_{\sigma=\uparrow,\downarrow} \int \frac{d^3 \mathbf{q}}{(2\pi)^D} \ln \left(1 + e^{-\beta E_q^{(\sigma)}} \right) - g n_\uparrow n_\downarrow \, ,$$

where the single-particle Hartree-Fock energy $E_q^{(\sigma)}$ given by

$$E_q^{(\sigma)} = \frac{\hbar^2 q^2}{2m} - \mu_\sigma + g n_{\sigma'}$$

and $\sigma' \neq \sigma$. At zero-temperature, after integration over momenta one finds the following internal energy density

$$\frac{E}{L^3} = \frac{\Omega}{L^3} + \mu_\uparrow n_\uparrow + \mu_\downarrow n_\downarrow = \frac{3}{5} \frac{\hbar^2}{2m} (6\pi^2)^{2/3} \left(n_\uparrow^{5/3} + n_\downarrow^{5/3} \right) + g n_\uparrow n_\downarrow \, ,$$

and chemical potentials

$$\mu_\sigma = \frac{\hbar^2}{2m} (6\pi^2)^{2/3} n_\sigma^{2/3} + g n_{\sigma'}$$

with $\sigma \neq \sigma'$. Given the Jacobian

$$J = \begin{pmatrix} \frac{\partial \mu_\uparrow}{\partial n_\uparrow} & \frac{\partial \mu_\uparrow}{\partial n_\downarrow} \\ \frac{\partial \mu_\downarrow}{\partial n_\uparrow} & \frac{\partial \mu_\downarrow}{\partial n_\downarrow} \end{pmatrix} = \begin{pmatrix} \frac{2}{3} \frac{\hbar^2}{2m} (6\pi^2)^{2/3} n_\uparrow^{-1/3} & g \\ g & \frac{2}{3} \frac{\hbar^2}{2m} (6\pi^2)^{2/3} n_\downarrow^{-1/3} \end{pmatrix}$$

the critical strength g_c of the unpolarized configuration is obtained setting $n_\uparrow = n_\downarrow = n/2$ and the determinant of J equal to zero. It follows

$$g_c = \frac{2}{3} \frac{\hbar^2}{2m} (6\pi^2)^{2/3} (\frac{n}{2})^{-1/3} \, .$$

Problem 9.3
Consider a two-component two-dimensional ($D = 2$) Fermi gas with repulsive contact interaction of strength g. Within the mean-field Hartree-Fock approximation, find the critical strength g_c above which the zero-temperature uniform unpolarized configuration becomes unstable (2D Stoner instability).

Solution

The procedure is quite similar to the one of the previous problem. At zero-temperature, after integration over momenta one finds the following internal energy density

$$\frac{E}{L^2} = \frac{\Omega}{L^2} + \mu_\uparrow n_\uparrow + \mu_\downarrow n_\downarrow = \pi \frac{\hbar^2}{m} \left(n_\uparrow^2 + n_\downarrow^2 \right) + g n_\uparrow n_\downarrow \,,$$

and chemical potentials

$$\mu_\sigma = 2\pi \frac{\hbar^2}{m} n_\sigma + g n_{\sigma'}$$

with $\sigma \neq \sigma'$. Also in this two-dimensional case, given the Jacobian

$$J = \begin{pmatrix} \frac{\partial \mu_\uparrow}{\partial n_\uparrow} & \frac{\partial \mu_\uparrow}{\partial n_\downarrow} \\ \frac{\partial \mu_\downarrow}{\partial n_\uparrow} & \frac{\partial \mu_\downarrow}{\partial n_\downarrow} \end{pmatrix} = \begin{pmatrix} 2\pi \frac{\hbar^2}{m} & g \\ g & 2\pi \frac{\hbar^2}{m} \end{pmatrix}$$

the critical strength g_c of the unpolarized configuration is obtained setting $n_\uparrow = n_\downarrow = n/2$ and the determinant of J equal to zero. It follows

$$g_c = 2\pi \frac{\hbar^2}{m} \,.$$

Problem 9.4
Consider a two-component one-dimensional ($D = 1$) Fermi gas with repulsive contact interaction of strength g. Within the mean-field Hartree-Fock approximation, find the critical strength g_c above which the zero-temperature uniform unpolarized configuration becomes unstable (1D Stoner instability).

Solution

The procedure is quite similar to the ones of the two previous problems. At zero-temperature, after integration over momenta one finds the following internal energy density

$$\frac{E}{L} = \frac{\Omega}{L} + \mu_\uparrow n_\uparrow + \mu_\downarrow n_\downarrow = \pi^2 \frac{\hbar^2}{6m} \left(n_\uparrow^3 + n_\downarrow^3 \right) + g n_\uparrow n_\downarrow \,,$$

and chemical potentials

$$\mu_\sigma = \pi^2 \frac{\hbar^2}{2m} n_\sigma^2 + g n_{\sigma'}$$

with $\sigma \neq \sigma'$. Also in this one-dimensional case, given the Jacobian

$$J = \begin{pmatrix} \frac{\partial \mu_\uparrow}{\partial n_\uparrow} & \frac{\partial \mu_\uparrow}{\partial n_\downarrow} \\ \frac{\partial \mu_\downarrow}{\partial n_\uparrow} & \frac{\partial \mu_\downarrow}{\partial n_\downarrow} \end{pmatrix} = \begin{pmatrix} \pi^2 \frac{\hbar^2}{m} n_\uparrow & g \\ g & \pi^2 \frac{\hbar^2}{m} n_\downarrow \end{pmatrix}$$

the critical strength g_c of the unpolarized configuration is obtained setting $n_\uparrow = n_\downarrow = n/2$ and the determinant of J equal to zero. It follows

$$g_c = \pi^2 \frac{\hbar^2}{m} \frac{n}{2} .$$

Problem 9.5

Prove that the fermionic partition function

$$\mathcal{Z} = \int \mathcal{D}[\psi_\sigma, \bar{\psi}_\sigma] \, \exp\left\{-\frac{1}{\hbar} S[\psi_\sigma, \bar{\psi}_\sigma]\right\} ,$$

where

$$S[\psi_\sigma, \bar{\psi}_\sigma] = \int_0^{\hbar\beta} d\tau \int d^3\mathbf{r} \left\{ \sum_{\sigma=\uparrow,\downarrow} \bar{\psi}_\sigma \left[\hbar\partial_\tau - \frac{\hbar^2}{2m}\nabla^2 - \mu\right]\psi_\sigma - g_0 \, \bar{\psi}_\uparrow \, \bar{\psi}_\downarrow \, \psi_\downarrow \, \psi_\uparrow \right\}$$

with $g_0 > 0$, can be rewritten (Hubbard-Stratonovich transformation) as

$$\mathcal{Z} = \mathcal{N} \int \mathcal{D}[\psi_\sigma, \bar{\psi}_\sigma] \mathcal{D}[\Delta, \Delta^*] \, \exp\left\{-\frac{1}{\hbar} S_e[\psi_\sigma, \bar{\psi}_\sigma, \Delta, \Delta^*]\right\} ,$$

where

$$S_e[\psi_\sigma, \bar{\psi}_\sigma, \Delta, \Delta^*] = \int_0^{\hbar\beta} d\tau \int d^3\mathbf{r} \left\{ \sum_{\sigma=\uparrow,\downarrow} \bar{\psi}_\sigma \left[\hbar\partial_\tau - \frac{\hbar^2}{2m}\nabla^2 - \mu\right]\psi_\sigma - \Delta^* \, \psi_\downarrow \, \psi_\uparrow \right.$$

$$\left. - \Delta \bar{\psi}_\uparrow \, \bar{\psi}_\downarrow + \frac{|\Delta|^2}{g_0} \right\} ,$$

$\Delta(\mathbf{r}, \tau)$ is a complex scalar field, and $\mathcal{N}$ is a normalization constant that disappears when one calculates thermal (and quantum) averages.

Solution

First of all we remark that in the BCS Lagrangian of Eq. (9.38) the complex order parameter Δ is constant and uniform, while here $\Delta(\mathbf{r}, \tau)$ is complex field of both space and time. Moreover, simply by inspection, it is easy to realize that one must indeed prove the following equation

$$\exp\left\{-\frac{1}{\hbar} \int_0^{\hbar\beta} d\tau \int d^3\mathbf{r} \left(-g_0 \, \bar{\psi}_\uparrow \, \bar{\psi}_\downarrow \, \psi_\downarrow \, \psi_\uparrow\right)\right\}$$

$$= \mathcal{N} \int \mathcal{D}[\Delta, \Delta^*] \, \exp\left\{-\frac{1}{\hbar} \int_0^{\hbar\beta} d\tau \int d^3\mathbf{r} \left(-\Delta^* \, \psi_\downarrow \, \psi_\uparrow - \Delta \bar{\psi}_\uparrow \, \bar{\psi}_\downarrow + \frac{|\Delta|^2}{g_0}\right)\right\} .$$

Now we prove the simple relation

$$e^{g_0 \bar{A} A} = \mathcal{N} \int \frac{dz dz^*}{2\pi i} e^{(z^* A + z \bar{A} - \frac{|z|^2}{g_0})} .$$

In fact, we observe that

$$g_0 \left| \frac{z}{g_0} - A \right|^2 = \frac{|z|^2}{g_0} - z^* A - z \bar{A} + g_0 \bar{A} A$$

and consequently

$$\int \frac{dz dz^*}{2\pi i} e^{(z^* A + z \bar{A} - \frac{|z|^2}{g_0})} = e^{g_0 \bar{A} A} \int \frac{dz dz^*}{2\pi i} e^{-g_0 |\frac{z}{g_0} - A|^2} = e^{g_0 \bar{A} A} g_0^2 \int \frac{d\zeta d\zeta^*}{2\pi i} e^{-g_0 |\zeta|^2} = e^{g_0 \bar{A} A} g_0 ,$$

setting $\zeta = z/g_0 - A$. It follows that, in this case, the normalization constant is $\mathcal{N} = 1/g_0$.

More generally, considering complex vectors $\mathbf{z} = (z_1, ..., z_N)$ and $\mathbf{A} = (A_1, ..., A_N)$ it is not difficult to prove that

$$e^{g_0 \bar{\mathbf{A}} \cdot \mathbf{A}} = \mathcal{N} \int \frac{d\mathbf{z} \cdot d\mathbf{z}^*}{2\pi i} e^{(\mathbf{z}^* \cdot \mathbf{A} + \mathbf{z} \cdot \bar{\mathbf{A}} - \frac{|z|^2}{g_0})}$$

with $\mathcal{N} = 1/g_0^N$. Finally, it is straightforward to extend this approach to complex fields obtaining what is required.

Further Reading

For the functional integration of the non-relativistic bosonic and fermionic fields:
N. Nagaosa, Quantum Field Theory in Condensed Matter Physics, Chap. 1, Sects. 1.1 and 1.2 (Springer, Berlin, 1999).
H.T.C. Stoof, K.B. Gubbels, and D.B.M. Dickerscheid, Ultracold Quantum Fields, Chap. 6, Sects. 6.1, 6.2, 6.3, and 6.4 (Springer, Berlin, 2009).
A. Altland and B. Simons, Condensed Matter Field Theory, Chap. 2, Sect. 2.1 (Cambridge Univ. Press, Cambridge, 2006).
For the dimensional regularization of Gaussian fluctuations in the case of interacting bosons and fermions:
L. Salasnich and F. Toigo, Zero-point energy of ultracold atoms, Phys. Rep. **640**, 1 (2016).
For the BCS theory in metals with various experimental data:
N.W. Ashcroft and N.D. Mermin, Solid State Physics (Holt-Sounders, Philadelphia, 1976).
J.P. Carbotte, Rev. Mod. Phys. **62**, 1027 (1990).
For the condensate fraction of superfluid fermions:
L. Salasnich, N. Manini, and A. Parola, Condensate Fraction of a Fermi Gas in the BCS-BEC Crossover, Phys. Rev. A **72**, 023621 (2005).

For a detailed account of the Ginzburg-Landau theory and its developments:

A. Larkin and A. Varlamov, Theory of Fluctuations in Superconductors (Oxford Univ. Press, Oxford, 2009).

For a comprehensive review on the state of the art in the study of the BCS-BEC crossover:

W. Zwerger (Ed.), The BCS-BEC Crossover and the Unitary Fermi Gas, Lecture Notes in Physics (Springer, Berlin, 2012).

Appendix A
Dirac Delta Function

In 1880 the self-taught electrical scientist Oliver Heaviside introduced the following function

$$\Theta(x) = \begin{cases} 1 \text{ for } x > 0 \\ 0 \text{ for } x < 0 \end{cases} \tag{A.1}$$

which is now called Heaviside step function. This is a discontinous function, with a discontinuity of first kind (jump) at $x = 0$, which is often used in the context of the analysis of electric signals. Moreover, it is important to stress that the Haviside step function appears also in the context of quantum statistical physics. In fact, the Fermi-Dirac function (or Fermi-Dirac distribution)

$$F_\beta(x) = \frac{1}{e^{\beta x} + 1} , \tag{A.2}$$

proposed in 1926 by Enrico Fermi and Paul Dirac to describe the quantum statistical distribution of electrons in metals, where $\beta = 1/(k_B T)$ is the inverse of the absolute temperature T (with k_B the Boltzmann constant) and $x = \epsilon - \mu$ is the energy ϵ of the electron with respect to the chemical potential μ, becomes the function $\Theta(-x)$ in the limit of very small temperature T, namely

$$\lim_{\beta \to +\infty} F_\beta(x) = \Theta(-x) = \begin{cases} 0 \text{ for } x > 0 \\ 1 \text{ for } x < 0 \end{cases} . \tag{A.3}$$

Inspired by the work of Heaviside, with the purpose of describing an extremely localized charge density, in 1930 Paul Dirac investigated the following "function"

$$\delta(x) = \begin{cases} +\infty \text{ for } x = 0 \\ 0 \quad \text{ for } x \neq 0 \end{cases} \tag{A.4}$$

imposing that

$$\int_{-\infty}^{+\infty} \delta(x) \, dx = 1 . \tag{A.5}$$

© Springer International Publishing AG 2017
L. Salasnich, *Quantum Physics of Light and Matter*, UNITEXT for Physics,
DOI 10.1007/978-3-319-52998-1

Unfortunately, this property of $\delta(x)$ is not compatible with the definition (A.4). In fact, from Eq. (A.4) it follows that the integral must be equal to zero. In other words, it does not exist a function $\delta(x)$ which satisfies both Eqs. (A.4) and (A.5). Dirac suggested that a way to circumvent this problem is to interpret the integral of Eq. (A.5) as

$$\int_{-\infty}^{+\infty} \delta(x)\, dx = \lim_{\epsilon \to 0^+} \int_{-\infty}^{+\infty} \delta_\epsilon(x)\, dx \, , \tag{A.6}$$

where $\delta_\epsilon(x)$ is a generic function of both x and ϵ such that

$$\lim_{\epsilon \to 0^+} \delta_\epsilon(x) = \begin{cases} +\infty & \text{for } x = 0 \\ 0 & \text{for } x \neq 0 \end{cases}, \tag{A.7}$$

$$\int_{-\infty}^{+\infty} \delta_\epsilon(x)\, dx = 1 \, . \tag{A.8}$$

Thus, the Dirac delta function $\delta(x)$ is a "generalized function" (but, strictly-speaking, not a function) which satisfy Eqs. (A.4) and (A.5) with the caveat that the integral in Eq. (A.5) must be interpreted according to Eq. (A.6) where the functions $\delta_\epsilon(x)$ satisfy Eqs. (A.7) and (A.8).

There are infinite functions $\delta_\epsilon(x)$ which satisfy Eqs. (A.7) and (A.8). Among them there is, for instance, the following Gaussian

$$\delta_\epsilon(x) = \frac{1}{\epsilon \sqrt{\pi}} e^{-x^2/\epsilon^2} \, , \tag{A.9}$$

which clearly satisfies Eq. (A.7) and whose integral is equal to 1 for any value of ϵ. Another example is the function

$$\delta_\epsilon(x) = \begin{cases} \frac{1}{\epsilon} & \text{for } |x| \leq \epsilon/2 \\ 0 & \text{for } |x| > \epsilon/2 \end{cases}, \tag{A.10}$$

which again satisfies Eq. (A.7) and whose integral is equal to 1 for any value of ϵ. In the following we shall use Eq. (A.10) to study the properties of the Dirac delta function.

According to the approach of Dirac, the integral involving $\delta(x)$ must be interpreted as the limit of the corresponding integral involving $\delta_\epsilon(x)$, namely

$$\int_{-\infty}^{+\infty} \delta(x)\, f(x)\, dx = \lim_{\epsilon \to 0^+} \int_{-\infty}^{+\infty} \delta_\epsilon(x)\, f(x)\, dx \, , \tag{A.11}$$

for any function $f(x)$. It is then easy to prove that

$$\int_{-\infty}^{+\infty} \delta(x)\, f(x)\, dx = f(0) \, . \tag{A.12}$$

by using Eq. (A.10) and the mean value theorem. Similarly one finds

$$\int_{-\infty}^{+\infty} \delta(x - c) f(x) \, dx = f(c) \, . \tag{A.13}$$

Several other properties of the Dirac delta function $\delta(x)$ follow from its definition. In particular

$$\delta(-x) = \delta(x) \, , \tag{A.14}$$

$$\delta(a \, x) = \frac{1}{|a|} \delta(x) \quad \text{with } a \neq 0 \, , \tag{A.15}$$

$$\delta(f(x)) = \sum_i \frac{1}{|f'(x_i)|} \delta(x - x_i) \quad \text{with } f(x_i) = 0 \, . \tag{A.16}$$

Up to now we have considered the Dirac delta function $\delta(x)$ with only one variable x. It is not difficult to define a Dirac delta function $\delta^{(D)}(\mathbf{r})$ in the case of a D-dimensional domain $\mathbb{R}^D$, where $\mathbf{r} = (x_1, x_2, ..., x_D) \in \mathbb{R}^D$ is a D-dimensional vector:

$$\delta^{(D)}(\mathbf{r}) = \begin{cases} +\infty & \text{for } \mathbf{r} = \mathbf{0} \\ 0 & \text{for } \mathbf{r} \neq \mathbf{0} \end{cases} \tag{A.17}$$

and

$$\int_{\mathbb{R}^D} \delta^{(D)}(\mathbf{r}) \, d^D\mathbf{r} = 1 \, . \tag{A.18}$$

Notice that sometimes $\delta^{(D)}(\mathbf{r})$ is written using the simpler notation $\delta(\mathbf{r})$. Clearly, also in this case one must interpret the integral of Eq. (A.18) as

$$\int_{\mathbb{R}^D} \delta^{(D)}(\mathbf{r}) \, d^D\mathbf{r} = \lim_{\epsilon \to 0^+} \int_{\mathbb{R}^D} \delta_\epsilon^{(D)}(\mathbf{r}) \, d^D\mathbf{r} \, , \tag{A.19}$$

where $\delta_\epsilon^{(D)}(\mathbf{r})$ is a generic function of both $\mathbf{r}$ and ϵ such that

$$\lim_{\epsilon \to 0^+} \delta_\epsilon^{(D)}(\mathbf{r}) = \begin{cases} +\infty & \text{for } \mathbf{r} = \mathbf{0} \\ 0 & \text{for } \mathbf{r} \neq \mathbf{0} \end{cases} , \tag{A.20}$$

$$\lim_{\epsilon \to 0^+} \int \delta_\epsilon^{(D)}(\mathbf{r}) \, d^D\mathbf{r} = 1 \, . \tag{A.21}$$

Several properties of $\delta(x)$ remain valid also for $\delta^{(D)}(\mathbf{r})$. Nevertheless, some properties of $\delta^{(D)}(\mathbf{r})$ depend on the space dimension D. For instance, one can prove the remarkable formula

$$\delta^{(D)}(\mathbf{r}) = \begin{cases} \frac{1}{2\pi} \nabla^2 (\ln |\mathbf{r}|) & \text{for } D = 2 \\ -\frac{1}{D(D-2)V_D} \nabla^2 \left(\frac{1}{|\mathbf{r}|^{D-2}} \right) & \text{for } D \geq 3 \end{cases} , \tag{A.22}$$

where $\nabla^2 = \frac{\partial^2}{\partial x_1^2} + \frac{\partial^2}{\partial x_2^2} + \cdots + \frac{\partial^2}{\partial x_D^2}$ and $V_D = \pi^{D/2}/\Gamma(1 + D/2)$ is the volume of a D-dimensional ipersphere of unitary radius, with $\Gamma(x)$ the Euler Gamma function. In the case $D = 3$ the previous formula becomes

$$\delta^{(3)}(\mathbf{r}) = -\frac{1}{4\pi}\nabla^2\left(\frac{1}{|\mathbf{r}|}\right) , \tag{A.23}$$

which can be used to transform the Gauss law of electromagnetism from its integral form to its differential form.

Appendix B
Fourier Transform

It was known from the times of Archimedes that, in some cases, the infinite sum of decreasing numbers can produce a finite result. But it was only in 1593 that the mathematician Francois Viete gave the first example of a function, $f(x) = 1/(1 - x)$, written as the infinite sum of power functions. This function is nothing else than the geometric series, given by

$$\frac{1}{1 - x} = \sum_{n=0}^{\infty} x^n , \quad \text{for } |x| < 1 . \tag{B.1}$$

In 1714 Brook Taylor suggested that any real function $f(x)$ which is infinitely differentiable in x_0 and sufficiently regular can be written as a series of powers, i.e.

$$f(x) = \sum_{n=0}^{\infty} c_n (x - x_0)^n , \tag{B.2}$$

where the coefficients c_n are given by

$$c_n = \frac{1}{n!} f^{(n)}(x_0) , \tag{B.3}$$

with $f^{(n)}(x)$ the n-th derivative of the function $f(x)$. The series (B.2) is now called Taylor series and becomes the so-called Maclaurin series if $x_0 = 0$. Clearly, the geometric series (B.1) is nothing else than the Maclaurin series, where $c_n = 1$. We observe that it is quite easy to prove the Taylor series: it is sufficient to suppose that Eq. (B.2) is valid and then to derive the coefficients c_n by calculating the derivatives of $f(x)$ at $x = x_0$; in this way one gets Eq. (B.3).

In 1807 Jean Baptiste Joseph Fourier, who was interested on wave propagation and periodic phenomena, found that any sufficiently regular real function function $f(x)$ which is periodic, i.e. such that

© Springer International Publishing AG 2017
L. Salasnich, *Quantum Physics of Light and Matter*, UNITEXT for Physics,
DOI 10.1007/978-3-319-52998-1

$$f(x + L) = f(x) , \tag{B.4}$$

where L is the periodicity, can be written as the infinite sum of sinusoidal functions, namely

$$f(x) = \frac{a_0}{2} + \sum_{n=1}^{\infty} \left[a_n \cos \left(n \frac{2\pi}{L} x \right) + b_n \sin \left(n \frac{2\pi}{L} x \right) \right] , \tag{B.5}$$

where

$$a_n = \frac{2}{L} \int_{-L/2}^{L/2} f(y) \cos \left(n \frac{2\pi}{L} y \right) dy , \tag{B.6}$$

$$b_n = \frac{2}{L} \int_{-L/2}^{L/2} f(y) \sin \left(n \frac{2\pi}{L} y \right) dy . \tag{B.7}$$

It is quite easy to prove also the series (B.5), which is now called Fourier series. In fact, it is sufficient to suppose that Eq. (B.5) is valid and then to derive the coefficients a_n and b_n by multiplying both side of Eq. (B.5) by $\cos \left(n \frac{2\pi}{L} x \right)$ and $\cos \left(n \frac{2\pi}{L} x \right)$ respectively and integrating over one period L; in this way one gets Eqs. (B.6) and (B.7).

It is important to stress that, in general, the real variable x of the function $f(x)$ can represent a space coordinate but also a time coordinate. In the former case L gives the spatial periodicity and $2\pi/L$ is the wavenumber, while in the latter case L is the time periodicity and $2\pi/L$ the angular frequency.

Taking into account the Euler formula

$$e^{i n \frac{2\pi}{L} x} = \cos \left(n \frac{2\pi}{L} x \right) + i \sin \left(n \frac{2\pi}{L} x \right) \tag{B.8}$$

with $i = \sqrt{-1}$ the imaginary unit, Fourier observed that his series (B.5) can be re-written in the very elegant form

$$f(x) = \sum_{n=-\infty}^{+\infty} f_n e^{i n \frac{2\pi}{L} x} , \tag{B.9}$$

where

$$f_n = \frac{1}{L} \int_{-L/2}^{L/2} f(y) e^{-i n \frac{2\pi}{L} y} dy \tag{B.10}$$

are complex coefficients, with $f_0 = a_0/2$, $f_n = (a_n - i b_n)/2$ if $n > 0$ and $f_n = (a_{-n} + i b_{-n})/2$ if $n < 0$, thus $f_n^* = f_{-n}$.

The complex representation (B.9) suggests that the function $f(x)$ can be periodic but complex, i.e. such that $f : \mathbb{R} \to \mathbb{C}$. Moreover, one can consider the limit $L \to +\infty$ of infinite periodicity, i.e. a function which is not periodic. In this limit Eq. (B.9) becomes the so-called Fourier integral (or Fourier anti-transform)

$$f(x) = \frac{1}{2\pi} \int_{-\infty}^{+\infty} \tilde{f}(k) \, e^{ikx} \, dk \tag{B.11}$$

with

$$\tilde{f}(k) = \int_{-\infty}^{\infty} f(y) \, e^{-iky} \, dy \tag{B.12}$$

the Fourier transform of $f(x)$. To prove Eqs. (B.11) and (B.12) we write Eq. (B.9) taking into account Eq. (B.10) and we find

$$f(x) = \sum_{n=-\infty}^{+\infty} \left(\frac{1}{L} \int_{-L/2}^{L/2} f(y) \, e^{-in\frac{2\pi}{L}y} \, dy \right) e^{in\frac{2\pi}{L}x} \,. \tag{B.13}$$

Setting

$$k_n = n\frac{2\pi}{L} \quad \text{and} \quad \Delta k = k_{n+1} - k_n = \frac{2\pi}{L} \tag{B.14}$$

the previous expression of $f(x)$ becomes

$$f(x) = \frac{1}{2\pi} \sum_{n=-\infty}^{+\infty} \left(\int_{-L/2}^{L/2} f(y) \, e^{-ik_n y} \, dy \right) e^{ik_n x} \, \Delta k \,. \tag{B.15}$$

In the limit $L \to +\infty$ one has $\Delta k \to dk$, $k_n \to k$ and consequently

$$f(x) = \frac{1}{2\pi} \int_{-\infty}^{+\infty} \left(\int_{-\infty}^{+\infty} f(y) \, e^{-iky} \, dy \right) e^{ikx} \, dk \,, \tag{B.16}$$

which gives exactly Eqs. (B.11) and (B.12). Note, however, that one gets the same result (B.16) if the Fourier integral and its Fourier transform are defined multiplying them respectively with a generic constant and its inverse. Thus, we have found that any sufficiently regular complex function $f(x)$ of real variable x which is globally integrable, i.e. such that

$$\int_{-\infty}^{+\infty} |f(x)| \, dx < +\infty \,, \tag{B.17}$$

can be considered as the (infinite) superposition of complex monocromatic waves e^{ikx}. The amplitude $\tilde{f}(k)$ of the monocromatic wave e^{ikx} is the Fourier transform of $f(x)$.

The Fourier transform $\tilde{f}(k)$ of a function $f(x)$ is sometimes denoted as $\mathcal{F}[f(x)](k)$, namely

$$\tilde{f}(k) = \mathcal{F}[f(x)](k) = \int_{-\infty}^{\infty} f(x) \, e^{-ikx} \, dx \,. \tag{B.18}$$

Table B.1 Fourier transforms $\mathcal{F}[f(x)](k)$ of simple functions $f(x)$, where $\delta(x)$ is the Dirac delta function, $sgn(x)$ is the sign function, and $\Theta(x)$ is the Heaviside step function

$f(x)$	$\mathcal{F}[f(x)](k)$
0	0
1	$2\pi\delta(k)$
$\delta(x)$	1
$\Theta(x)$	$\frac{1}{ik} + \pi\,\delta(k)$
$e^{ik_0 x}$	$2\pi\,\delta(k - k_0)$
$e^{-x^2/(2a^2)}$	$a\sqrt{2\pi}e^{-a^2k^2/2}$
$e^{-a|x|}$	$\frac{2a}{a^2+k^2}$
$sgn(x)$	$\frac{2}{ik}$
$\sin(k_0 x)$	$\frac{\pi}{i}[\delta(k - k_0) - \delta(k + k_0)]$
$\cos(k_0 x)$	$\pi[\delta(k - k_0) + \delta(k + k_0)]$

The Fourier transform $\mathcal{F}[f(x)](k)$ has many interesting properties. For instance, due to the linearity of the integral the Fourier transform is clearly a linear map:

$$\mathcal{F}[a\,f(x) + b\,g(x)](k) = a\,\mathcal{F}[f(x)](k) + b\,\mathcal{F}[g(x)](k) . \tag{B.19}$$

Moreover, one finds immediately that

$$\mathcal{F}[f(x - a)](k) = e^{-ika}\,\mathcal{F}[f(x)](k) , \tag{B.20}$$
$$\mathcal{F}[e^{ik_0 x} f(x)](k) = \mathcal{F}[f(x)](k - k_0) . \tag{B.21}$$
$$\mathcal{F}[x\,f(x)](k) = i\,\tilde{f}'(k) , \tag{B.22}$$
$$\mathcal{F}[f^{(n)}(x)](k) = (ik)^n\,\tilde{f}(k) , \tag{B.23}$$

where $f^{(n)}(x)$ is the n-th derivative of $f(x)$ with respect to x (Table B.1).

In the Table we report the Fourier transforms $\mathcal{F}[f(x)](k)$ of some elementary functions $f(x)$, including the Dirac delta function $\delta(x)$ and the Heaviside step function $\Theta(x)$. We insert also the sign function $sgn(x)$ defined as: $sgn(x) = 1$ for $x > 0$ and $sgn(x) = -1$ for $x < 0$. The table of Fourier transforms clearly shows that the Fourier transform localizes functions which is delocalized, while it delocalizes functions which are localized. In fact, the Fourier transform of a constant is a Dirac delta function while the Fourier transform of a Dirac delta function is a constant. In general, it holds the following uncertainty theorem

$$\Delta x\,\Delta k \geq \frac{1}{2} , \tag{B.24}$$

where

$$\Delta x^2 = \int_{-\infty}^{\infty} x^2\,|f(x)|^2\,dx - \left(\int_{-\infty}^{\infty} x\,|f(x)|^2\,dx\right)^2 \tag{B.25}$$

and

$$\Delta k^2 = \int_{-\infty}^{\infty} k^2 \, |\tilde{f}(k)|^2 \, dk - \left(\int_{-\infty}^{\infty} k \, |\tilde{f}(k)|^2 \, dk \right)^2 \tag{B.26}$$

are the spreads of the wavepackets respectively in the space x and in the dual space k. This theorem is nothing else than the uncertainty principle of quantum mechanics formulated by Werner Heisenberg in 1927, where x is the position and k is the wavenumber. Another interesting and intuitive relationship is the Parseval identity, given by

$$\int_{-\infty}^{+\infty} |f(x)|^2 dx = \int_{-\infty}^{+\infty} |\tilde{f}(k)|^2 dk . \tag{B.27}$$

It is important to stress that the power series, the Fourier series, and the Fourier integral are special cases of the quite general expansion

$$f(x) = \sum_{\alpha} \!\!\!\!\!\int f_\alpha \, \phi_\alpha(x) \, d\alpha \tag{B.28}$$

of a generic function $f(x)$ in terms of a set $\phi_\alpha(x)$ of basis functions spanned by the parameter α, which can be a discrete or a continuous variable. A large part of modern mathematical analysis is devoted to the study of Eq. (B.28) and its generalization.

The Fourier transform is often used in electronics. In that field of research the signal of amplitude f depends on time t, i.e. $f = f(t)$. In this case the dual variable of time t is the frequency ω and the fourier integral is usually written as

$$f(t) = \frac{1}{2\pi} \int_{-\infty}^{+\infty} \tilde{f}(\omega) \, e^{-i\omega t} \, dk \tag{B.29}$$

with

$$\tilde{f}(\omega) = \mathcal{F}[f(t)](\omega) = \int_{-\infty}^{\infty} f(t) \, e^{i\omega t} \, dt \tag{B.30}$$

the Fourier transform of $f(t)$. Clearly, the function $f(t)$ can be seen as the Fourier anti-transform of $\tilde{f}(\omega)$, in symbols

$$f(t) = \mathcal{F}^{-1}[\tilde{f}(\omega)](t) = \mathcal{F}^{-1}[\mathcal{F}[f(t)](\omega)](t) , \tag{B.31}$$

which obviously means that the composition $\mathcal{F}^{-1} \circ \mathcal{F}$ gives the identity.

More generally, if the signal f depends on the 3 spatial coordinates $\mathbf{r} = (x, y, z)$ and time t, i.e. $f = f(\mathbf{r}, t)$, one can introduce Fourier transforms from $\mathbf{r}$ to $\mathbf{k}$, from t to ω, or both. In this latter case one obviously obtains

$$f(\mathbf{r}, t) = \frac{1}{(2\pi)^4} \int_{\mathbb{R}^4} \tilde{f}(\mathbf{k}, \omega) \, e^{i(\mathbf{k} \cdot \mathbf{x} - \omega t)} \, d^3\mathbf{k} \, d\omega \tag{B.32}$$

with

$$\tilde{f}(\mathbf{k}, \omega) = \mathcal{F}[f(\mathbf{r}, t)](\mathbf{k}, \omega) = \int_{\mathbb{R}^4} f(\mathbf{r}, t) \, e^{-i(\mathbf{k}\cdot\mathbf{r}-\omega t)} \, d^3\mathbf{r} \, dt \ . \qquad (B.33)$$

Also in this general case the function $f(\mathbf{r}, t)$ can be seen as the Fourier anti-transform of $\tilde{f}(\mathbf{k}, \omega)$, in symbols

$$f(\mathbf{r}, t) = \mathcal{F}^{-1}[\tilde{f}(\mathbf{k}, \omega)](\mathbf{r}, t) = \mathcal{F}^{-1}[\mathcal{F}[f(\mathbf{r}, t)](\mathbf{k}, \omega)](\mathbf{r}, t) \ . \qquad (B.34)$$

Appendix C
Laplace Transform

The Laplace transform is an integral transformation, similar but not equal to the Fourier transform, introduced in 1737 by Leonard Euler and independently in 1782 by Pierre-Simon de Laplace. Nowaday the Laplace transform is mainly used to solve non-homogeneous ordinary differential equations with constant coefficients.

Given a sufficiently regular function $f(t)$ of time t, the Laplace transform of $f(t)$ is the function $F(s)$ such that

$$F(s) = \int_0^{+\infty} f(t)\, e^{-st}\, dt \ , \qquad (C.1)$$

where s is a complex number. Usually the integral converges if the real part $Re(s)$ of the complex number s is greater than critical real number x_c, which is called abscissa of convergence and unfortunately depends on $f(t)$. The Laplace transform $F(s)$ of a function $f(t)$ is sometimes denoted as $\mathcal{L}[f(t)](s)$, namely

$$F(s) = \mathcal{L}[f(t)](s) = \int_0^{\infty} f(t)\, e^{-st}\, dt \ . \qquad (C.2)$$

For the sake of completeness and clarity, we write also the Fourier transform $\tilde{f}(\omega)$, denoted as $\mathcal{F}[f(t)](\omega)$, of the same function

$$\tilde{f}(\omega) = \mathcal{F}[f(t)](\omega) = \int_{-\infty}^{\infty} f(t)\, e^{i\omega t}\, dt \ . \qquad (C.3)$$

First of all we notice that the Laplace transform depends on the behavior of $f(t)$ for non negative values of t, while the Fourier transform depends also on the behavior of $f(t)$ for negative values of t. This is however not a big problem, because we can set $f(t) = 0$ for $t < 0$ (or equivalently we can multiply $f(t)$ by the Heaviside step function $\Theta(t)$), and then the Fourier transform becomes

© Springer International Publishing AG 2017
L. Salasnich, *Quantum Physics of Light and Matter*, UNITEXT for Physics,
DOI 10.1007/978-3-319-52998-1

Table C.1 Laplace transforms $\mathcal{L}[f(t)](s)$ of simple functions $f(t)$, where $\delta(t)$ is the Dirac delta function and $\Theta(t)$ is the Heaviside step function, and $\tau > 0$ and n positive integer

$f(t)$	$\mathcal{L}[f(t)](s)$
0	0
1	$\frac{1}{s}$
$\delta(t-\tau)$	$e^{-\tau s}$
$\Theta(t-\tau)$	$\frac{e^{-\tau s}}{s}$
t^n	$\frac{n!}{s^{n+1}}$
e^{-at}	$\frac{1}{s+a}$
$e^{-a|t|}$	$\frac{2a}{a^2-s^2}$
$\sin(at)$	$\frac{a}{s^2+a^2}$
$\cos(at)$	$\frac{s}{s^2+a^2}$

$$\tilde{f}(\omega) = \mathcal{F}[f(t)](\omega) = \int_0^\infty f(t)\,e^{i\omega t}\,dt \ . \tag{C.4}$$

Moreover, it is important to observe that, comparing Eq. (C.2) with Eq. (C.4), if both $\mathcal{F}$ and $\mathcal{L}$ of $f(t)$ exist, we obtain

$$F(s) = \mathcal{L}[f(t)](s) = \mathcal{F}[f(t)](is) = \tilde{f}(is) \ , \tag{C.5}$$

or equivalently

$$\tilde{f}(\omega) = \mathcal{F}[f(t)](\omega) = \mathcal{L}[f(t)](-i\omega) = F(-i\omega) \ . \tag{C.6}$$

Remember that ω is a real variable while s is a complex variable. Thus we have found that for a generic function $f(t)$, such that $f(t) = 0$ for $t < 0$, the Laplace transform $F(s)$ and the Fourier transform $\tilde{f}(\omega)$ are simply related to each other.

In the following Table we report the Laplace transforms $\mathcal{L}[f(t)](s)$ of some elementary functions $f(t)$, including the Dirac delta function $\delta(t)$ and the Heaviside step function $\Theta(t)$, forgetting about the possible problems of regularity (Table C.1).

We now show that writing $f(t)$ as the Fourier anti-transform of $\tilde{f}(\omega)$ one can deduce the formula of the Laplace anti-transform of $F(s)$. In fact, one has

$$f(t) = \mathcal{F}^{-1}[\tilde{f}(\omega)](t) = \frac{1}{2\pi}\int_{-\infty}^\infty \tilde{f}(\omega)\,e^{-i\omega t}\,d\omega \ . \tag{C.7}$$

Because $\tilde{f}(\omega) = F(-i\omega)$ one finds also

$$f(t) = \frac{1}{2\pi}\int_{-\infty}^\infty F(-i\omega)\,e^{-i\omega t}\,d\omega \ . \tag{C.8}$$

Using $s = -i\omega$ as integration variable, this integral representation becomes

$$f(t) = \frac{1}{2\pi i} \int_{-i\infty}^{+i\infty} F(s)\, e^{st}\, ds \,, \tag{C.9}$$

where the integration is now a contour integral along any path γ in the complex plane of the variable s, which starts at $s = -i\infty$ and ends at $s = i\infty$. What we have found is exactly the Laplace anti-transform of $F(s)$, in symbols

$$f(t) = \mathcal{L}^{-1}[F(s)](t) = \frac{1}{2\pi i} \int_{-i\infty}^{+i\infty} F(s)\, e^{st}\, ds \,. \tag{C.10}$$

Remember that this function is such that $f(t) = 0$ for $t < 0$. The fact that the function $f(t)$ is the Laplace anti-transform of $F(s)$ can be symbolized by

$$f(t) = \mathcal{L}^{-1}[F(s)](t) = \mathcal{L}^{-1}[\mathcal{L}[f(t)](s)](t) \,, \tag{C.11}$$

which means that the composition $\mathcal{L}^{-1} \circ \mathcal{L}$ gives the identity.

The Laplace transform $\mathcal{L}[f(t)](s)$ has many interesting properties. For instance, due to the linearity of the integral the Laplace transform is clearly a linear map:

$$\mathcal{L}[a\, f(t) + b\, g(t)](s) = a\, \mathcal{L}[f(t)](s) + b\, \mathcal{L}[g(t)](s) \,. \tag{C.12}$$

Moreover, one finds immediately that

$$\mathcal{L}[f(t-a)\Theta(t-a)](s) = e^{-as}\, \mathcal{L}[f(t)](s) \,, \tag{C.13}$$

$$\mathcal{L}[e^{at} f(t)](s) = \mathcal{L}[f(t)](s-a) \,. \tag{C.14}$$

For the solution of non-homogeneous ordinary differential equations with constant coefficients, the most important property of the Laplace transform is the following

$$\mathcal{L}[f^{(n)}(t)](s) = s^n\, \mathcal{L}[f(t)](s) - s^{n-1} f(0) - s^{n-2} f^{(1)}(0) - \cdots - s\, f^{(n-2)} - f^{(n-1)}(0) \tag{C.15}$$

where $f^{(n)}(t)$ is the n-th derivative of $f(t)$ with respect to t. For instance, in the simple cases $n = 1$ and $n = 2$ one has

$$\mathcal{L}[f'(t)](s) = s\, F(s) - f(0) \,, \tag{C.16}$$

$$\mathcal{L}[f''(t)](s) = s^2\, F(s) - s f'(0) - f(0) \,, \tag{C.17}$$

by using $F(s) = \mathcal{L}[f(t)](s)$. The proof of Eq. (C.15) is straightforward performing integration by parts. Let us prove, for instance, Eq. (C.16):

$$\mathcal{L}[f'(t)](s) = \int_0^{+\infty} f'(t)\, e^{-st}\, dt = \left[f(t)\, e^{-st} \right]_0^{+\infty} - \int_0^{+\infty} f(t)\, \frac{d}{dt}\left(e^{-st} \right) dt$$

$$= -f(0) + s \int_0^{+\infty} f(t)\, e^{-st}\, dt = -f(0) + s\, \mathcal{L}[f(t)](s) \quad \text{(C.18)}$$

$$= -f(0) + s\, F(s)\, .$$

We now give a simple example of the Laplace method to solve ordinary differential equations by considering the differential problem

$$f'(t) + f(t) = 1 \quad \text{with} \quad f(0) = 2\, . \tag{C.19}$$

We apply the Laplace transform to both sides of the differential problem

$$\mathcal{L}[f'(t) + f(t)](s) = \mathcal{L}[1](s) \tag{C.20}$$

obtaining

$$s\, F(s) - 2 + F(s) = \frac{1}{s}\, . \tag{C.21}$$

This is now an algebric problem with solution

$$F(s) = \frac{1}{s(s+1)} + \frac{2}{s+1} = \frac{1}{s} - \frac{1}{s+1} + \frac{2}{s+1} = \frac{1}{s} + \frac{1}{s+1}\, . \tag{C.22}$$

Finally, we apply the Laplace anti-transform

$$f(t) = \mathcal{L}^{-1}\left[\frac{1}{s} + \frac{1}{s+1} \right](t) = \mathcal{L}^{-1}\left[\frac{1}{s} \right](t) + \mathcal{L}^{-1}\left[\frac{1}{s+1} \right](t)\, . \tag{C.23}$$

By using our Table of Laplace transforms we find immediately the solution

$$f(t) = 1 + e^{-t}\, . \tag{C.24}$$

The Laplace transform can be used to solve also integral equations. In fact, one finds

$$\mathcal{L}\left[\int_0^t f(y)\, dy \right](s) = \frac{1}{s} F(s)\, , \tag{C.25}$$

and more generally

$$\mathcal{L}\left[\int_{-\infty}^t f(y)\, g(t - y)\, dy \right](s) = \frac{I_0}{s} + F(s)\, G(s)\, , \tag{C.26}$$

where

$$I_0 = \int_{-\infty}^{0} f(y)\, g(t-y)\, dy \; . \tag{C.27}$$

Notice that the integral which appears in Eq. (C.25) is, by definition, the convolution $(f * g)(t)$ of two functions, i.e.

$$(f * g)(t) = \int_{0}^{t} f(y)\, g(t-y)\, dy \; . \tag{C.28}$$

Bibliography

1. F.T. Arecchi, Measurement of the statistical distribution of Gaussian and laser sources. Phys. Rev. Lett. **15**, 912 (1965)
2. A. Altland, B. Simons, *Condensed Matter Field Theory* (Cambridge University Press, Cambridge, 2006)
3. V.B. Berestetskii, E.M. Lifshitz, L.P. Pitaevskii, *Relativistic Quantum Theory*, vol. 4 of Course of Theoretical Physics (Pergamon Press, Oxford, 1971)
4. J.D. Bjorken, S.D. Drell, *Relativistic Quantum Mechanics* (McGraw-Hill, New York, 1964)
5. B.H. Bransden, C.J. Joachain, *Physics of Atoms and Molecules* (Prentice Hall, Upper Saddle River, 2003)
6. G. Bressi, G. Carugno, R. Onofrio, G. Ruoso, Measurement of the Casimir force between parallel metallic surfaces. Phys. Rev. Lett. **88**, 041804 (2002)
7. J.P. Carbotte, Rev. Mod. Phys. **62**, 1027 (1990)
8. C. Cohen-Tannoudji, B. Dui, F. Laloe, *Quantum Machanics* (Wiley, New York, 1991)
9. P.A.M. Dirac, *The Principles of Quantum Mechanics* (Oxford University Press, Oxford, 1982)
10. A.L. Fetter, J.D. Walecka, *Quantum Theory of Many-Particle Systems* (Dover Publications, New York, 2003)
11. N.W. Ashcroft, N.D. Mermin, *Solid State Physics* (Holt-Sounders, Philadelphia, 1976)
12. K. Huang, *Statistical Mechanics* (Wiley, New York, 1987)
13. S.K. Lamoreaux, Demonstration of the Casimir force in the 0.6 to 6 m range. Phys. Rev. Lett. **78**, 5 (1997)
14. A. Larkin, A. Varlamov, *Theory of Fluctuations in Superconductors* (Oxford University Press, Oxford, 2009)
15. M. Le Bellac, *A Short Introduction to Quantum Information and Quantum Computation* (Cambridge University Press, Cambridge, 2006)
16. U. Leonhardt, *Measuring the Quantum State of Light* (Cambridge University Press, Cambridge, 1997)
17. E. Lipparini, *Modern Many-Particle Physics: Atomic Gases, Quantum Dots and Quantum Fluids* (World Scientific, Singapore, 2003)
18. F. Mandl, G. Shaw, *Quantum Field Theory* (Wiley, New York, 1984)
19. G. Mazzarella, L. Salasnich, A. Parola, F. Toigo, Coherence and entanglement in the ground-state of a bosonic Josephson junction:from macroscopic Schrdinger cats to separable Fock states. Phys. Rev. A **83**, 053607 (2011)
20. N. Nagaosa, *Quantum Field Theory in Condensed Matter Physics* (Springer, Berlin, 1999)
21. W. Rindler, *Introduction to Special Relativity* (Oxford University Press, Oxford, 1991)
22. R.W. Robinett, *Quantum Mechanics: Classical Results, Modern Systems, and Visualized Examples* (Oxford University Press, Oxford, 2006)
23. L. Salasnich, N. Manini, A. Parola, Condensate fraction of a Fermi gas in the BCS-BEC crossover. Phys. Rev. A **72**, 023621 (2005)

24. L. Salasnich, F. Toigo, Zero-point energy of ultracold atoms. Phys. Rep. **640**, 1 (2016)
25. A. Scott (ed.), *Encyclopedia of Nonlinear Science* (Taylor and Francis, New York, 2005)
26. M.O. Scully, M.S. Zubairy, *Quantum Optics* (Cambridge University Press, Cambridge, 1997)
27. H.T.C. Stoof, K.B. Gubbels, D.B.M. Dickerscheid, *Ultracold Quantum Fields* (Springer, Berlin, 2009)
28. W. Zwerger (ed.), *The BCS-BEC Crossover and the Unitary Fermi Gas*, Lecture Notes in Physics (Springer, Berlin, 2012)

Printed in the United States
By Bookmasters